［英］ 詹姆斯·摩尔 保罗·尼禄◎著

James Moore & Paul Nero

遗失的创意

上册

陈明晖 孙怡 刘晓燕◎译

陈新忠 王俊英◎注

刘夙 王茜◎校订

科学普及出版社

·北京·

图书在版编目（CIP）数据

遗失的创意. 上册：汉、英 /（英）詹姆斯·摩尔，（英）保罗·尼禄著；陈明晖，孙怡，刘晓燕译. —北京：科学普及出版社，2021.1
书名原文：Pigeon Guided Missiles
ISBN 978-7-110-10011-0

Ⅰ．①遗… Ⅱ．①詹… ②保… ③陈… ④孙… ⑤刘… Ⅲ．①创造发明—青少年读物—汉、英 Ⅳ．①N19-49

中国版本图书馆 CIP 数据核字（2019）第225667号

策划编辑	郑洪炜
责任编辑	郑洪炜
封面设计	逸水翔天
正文设计	逸水翔天
责任校对	焦　宁
责任印制	马宇晨

出　　版	科学普及出版社
发　　行	中国科学技术出版社有限公司发行部
地　　址	北京市海淀区中关村南大街 16 号
邮　　编	100081
发行电话	010-62173865
传　　真	010-62173081
网　　址	http://www.cspbooks.com.cn

开　　本	787mm×1092mm　1/16
字　　数	580千字
印　　张	33.25
印　　数	1—5000册
版　　次	2021年1月第1版
印　　次	2021年1月第1次印刷
印　　刷	三河市荣展印务有限公司
书　　号	ISBN 978-7-110-10011-0/N·249
定　　价	86.00元（全2册）

CONTENTS
目　录

OUR THANKS

注释

① livelihood ['laɪvlihʊd] *n.* 谋生之道，营生
② mundane [mʌn'deɪn] *adj.* 平凡的，单调的
③ splendid ['splendɪd] *adj.* 值得高度赞赏的，光辉的

Like all writers, we earn our livelihoods① not just through the words we put on the page, but on ideas; some big, some small, a few exciting, others more mundane②. Occasionally we go through a purple patch of rather splendid③ ideas; more frequently our concepts fall by the wayside, perhaps even laughed at by our clients and families and friends. Or worse still, ignored. That's when we know we've really got things wrong. Now if we were adept at coming up with questionable ideas, surely it must happen to others? With this germ of a concept, the origins of Pigeon-Guided Missiles were formed. But we couldn't have done it alone. For making sense of some subjects in which we were hardly experts, for sourcing pictures where we had none, and for the occasional sanity or fact check, we needed help.

So we'd like to extend our warmest thanks to: Linda Bailey, Malcolm Barres-Baker, Tim Clarke, Jason Flahardy, Professor Christian G. Fritz, Amy Frost, Harry Gates, Stephen Guy, Simon

致　谢

　　像所有作家一样，我们不仅靠诉诸纸上的文字谋生，还得靠一些或大或小的想法糊口，这些想法有的动人心弦，但更多的则平淡无奇。我们有时可能会想到一些极妙的主意，但大多数情况下这些想法往往半途而废，甚至还有可能遭到亲朋好友的嘲笑，最糟糕的情况是别人对此不屑一顾。这时我们才真正知道自己想错了。既然我们是如此擅长提出不靠谱的想法，是不是别人也一定如此呢？受到这个主意的启发，我们产生了写作《遗失的创意》这本书的想法。但是我们不能独自完成这一任务，为了弄明白一些我们不在行的问题，为了找到一些我们无法找到的图片以及偶尔查证观点或查证事实，我们需要大家的帮助。

　　因此，我们要衷心感谢：Linda Bailey、Malcolm Barres-Baker、Tim Clarke、Jason Flahardy、Christian G. Fritz教授、Amy Frost、Harry Gates、Stephen Guy、Simon Hamlet、Jan Hebditch、Richard Hebditch、Patricia Hansen、Ted Huetter、Aideen Jenkins、Owen

Hamlet, Jan Hebditch, Richard Hebditch, Patricia Hansen, Ted Huetter, Aideen Jenkins, Owen Jones, Matthew Marshall, Amanda McNally, Geoff Moore, Laurie Moore, Philippa Moore, Sam Moore, Tamsin Moore, Dr Tom Moore, Marguerite Moran, Dr Claire Nesbitt, Neil Paterson, William Poole, Hannah Reyonlds, Kasey Richter, Sarah Sarkhel, David Smith, Robert Smith, Peter Spurgeon, Mia Sykes, Julie S. Vargas, Spencer Vignes, Paul Waddington and Dean Weber.

Thank you all.

Jones、Matthew Marshall、Amanda McNally、Geoff Moore、Laurie Moore、Philippa Moore、Sam Moore、Tamsin Moore、Tom Moore博士、Marguerite Moran、Claire Nesbitt博士、Neil Paterson、William Poole、Hannah Reyonlds、Kasey Richter、Sarah Sarkhel、David Smith、Robert Smith、Peter Spurgeon、Mia Sykes、Julie S. Vargas、Spencer Vignes、Paul Waddington 和 Dean Weber。

感谢各位!

INTRODUCTION

IN SEARCH OF HISTORY'S LOST IDEAS

If at first the idea is not absurd[1], then there is no hope for it.

Albert Einstein

Success is going from failure to failure without losing enthusiasm.

Winston Churchill

The past is littered with examples of grandiose[2] schemes [3] and ambitious ideas that never quite took off. While there are plenty of books about the visions, plans and inventions that did go on to transform our world, these heroic 'might have beens' have found themselves largely forgotten, carelessly dumped[4] on the scrapheap[5] of history.

This book sets out to rescue some of those incredible concepts and dreams, which, however briefly, promised to change our lives

注释

① absurd [əb'sɜ:d] *adj.* 荒谬的，荒唐的
② grandiose ['grændiəʊs] *adj.* 宏伟的，壮丽的
③ scheme [ski:m] *n.* 计划，规划，方案
④ dump [dʌmp] *vt.* 倾倒，丢下
⑤ scrapheap ['skræphi:p] *n.* 废料堆，废物堆

追寻遗失的创意

如果一个想法最初听起来并不荒谬，那就不要对它寄予太大希望了。

——阿尔伯特·爱因斯坦

所谓成功，就是经历无数次失败，但始终保持热情。

——温斯顿·丘吉尔

回首往昔，奇思妙想众多，未实现者亦不少。关于那些改变世界的想法、计划和发明的书，简直是汗牛充栋，但那些曾经可能改变人类世界的大胆想法、计划和发明，却大多为人们所遗忘，被漫不经心地扔进了历史的废纸堆。

本书旨在重新找回这些曾经有可能改变人类生活和地球面貌的伟大创意和梦想，尽管它们每个都很"短命"。此外，本书还展现了这些创意和梦想背后的那些激情四射、

注释

① flawed [flɔ:d] *adj.* 有缺点的

② edifice ['edɪfɪs] *n.* （巨大而宏伟的）建筑物，大厦

③ outlandish [aʊt'lændɪʃ] *adj.* 稀奇古怪的，奇特的

④ eccentric [ɪk'sentrɪk] *adj.* 古怪的，怪癖的，异乎寻常的

⑤ folly ['fɒli] *n.* 愚行，蠢笨

⑥ blunder ['blʌndə(r)] *n.* （因无知、粗心等造成）的错误，疏忽

⑦ precarious [prɪ'keəriəs] *adj.* 不确定的

⑧ serendipity [ˌserən'dɪpəti] *n.* 侥幸发现的东西，意外的发现

⑨ testament ['testəmənt] *n.* 证据，证明

⑩ conceive [kən'si:v] *vt.& vi.* 想象，设想

⑪ endeavor [ɪn'devə] *vt.& vi.* 尽力，竭力

⑫ proponent [prə'pəunənt] *n.* 支持者，拥护者

⑬ radical ['rædɪkl] *adj.* 激进的

⑭ rival ['raɪvl] *n.* 对手，竞争者

⑮ gothic ['gɒθɪk] *adj.* 哥特风格的

⑯ succumb [sə'kʌm] *vi.* 屈从，舍弃

⑰ Trafalgar Square 特拉法尔加广场

⑱ audacious [ɔ:'deɪʃəs] *adj.* 大胆的，敢于冒险的

and the face of our planet. It also reveals the fascinating, flawed① and tirelessly optimistic characters behind them. From soaring edifices② that never were, to fanciful devices to change our daily lives; some of the ideas are simple, others breathtakingly outlandish③. Eccentric④ engines of war, peculiar methods of transport, sporting follies⑤ and nation-building blunders⑥ – they're all here. Each example proves the precarious⑦ nature of success and shows how, but for a bit of serendipity⑧, the world we live in today could have been very different. They are also testament⑨ to the dedication, inspiration and, at times, sheer bloody-mindedness of the people who conceived⑩ them.

Pigeon-Guided Missiles: and 49 Other Ideas that Never Took Off covers a lot of ground and we don't pretend to have been scientific in our choices. But in choosing these stories, we've relied on a few guiding principles. Every chapter endeavours⑪ to reveal a relatively unknown proposition from history, investigates what drove its proponents⑫ and why they failed. We have looked for engaging tales that are often humorous and sometimes tragic, but always contain the seeds of truly radical⑬ thinking.

In this book we'll discover that some, like Sir Edward Watkin's attempt to build a rival⑭ to the Eiffel Tower in London, or William Beckford's enormous gothic⑮ home, were the victim of bad planning. Others, such as Fulton's flying car, succumbed⑯ to lack of cash. The pyramid in London's Trafalgar Square⑰ and Bessemer's ship to cure seasickness were simply too audacious⑱, while the first gas-powered traffic lights were

满身缺点、不知疲倦、乐观向上的人物。从亘古未有的高楼大厦到改变日常生活的古怪装置，一些想法比较简单，一些想法则相当奇异。古怪的战争工具、运输方法、体育的方面愚蠢想法和国家建筑大错在本书中应有尽有。每个故事都证明了成功的不确定性，表明一点意外就可能导致我们今天生活的世界变得大不相同。它们还是这些人的奉献、灵感和存心"找茬"的证据。

本书范畴广泛，而我们在选择书中的故事时并未刻意讲究科学。不过在选择这些故事时，我们确实以一些指导原则为依据。各章致力于展现历史中相对鲜为人知的计划，研究推动其诞生和失败的原因。我们不断在寻找精彩的故事，这些故事往往十分幽默，有时甚至非常悲惨，但常常孕育着真正变革性思考的种子。

在本书中，我们会发现一些想法是不良设计的牺牲品，比如爱德华·沃特金爵士尝试在伦敦修建与埃菲尔铁塔相媲美的建筑或威廉·贝克福德的巨大哥特式房屋，另一些想法则是由于缺乏资金而夭折的，比如富尔顿的飞行汽车。伦敦特拉法尔加广场的金字塔和治疗晕船的"贝西默号"轮船太过大胆，而第一款气动红绿灯又太超前。水泥雷达式预警盘由于技术进步而被淘汰，罗伯特·弗拉德的永动机和哈里·格林德尔·马修斯的死亡射线似乎只是由于违背了物理定律而失败。像许多好的想法一样，一些想法运气不佳，未引起重要人物的关注，在当时盛行的政治气候或商业环境中被认为荒谬至极或被扔进了垃圾堆。

通过我们的研究，有一样东西变得非常清晰，那

注释

① perpetual motion machine 永动机
② defy [dɪˈfaɪ] vt. （公然）违抗，反抗
③ preposterous [prɪˈpɒstərəs] adj. 荒谬可笑的
④ prevailing [prɪˈveɪlɪŋ] adj. 盛行的
⑤ industrious [ɪnˈdʌstriəs] adj. 勤奋的，勤恳的
⑥ verge [vɜːdʒ] vi. 接近，濒临
⑦ obsessive [əbˈsesɪv] adj. 使人着迷的

too ahead of their time. Radar-style warning dishes made from concrete were overtaken by technological developments. Robert Fludd's perpetual motion machine① and Harry Grindell Matthews' death ray simply seemed to defy② the laws of physics. A lot, like so many good ideas, were just unlucky; failing to catch the eye of those who mattered, dismissed as preposterous③ or left gathering dust thanks to the prevailing④ political or commercial climate.

One thing becomes very clear from our research – that history is so often driven by a few industrious⑤ men and women, who, verging⑥ on the obsessive⑦, never stop coming up with ideas that they genuinely believe will be a step forward for science, will help mankind or enhance the world we live in. As this book shows, they don't always get it right, but the story of how they fail, often spectacularly, is endlessly captivating and there is usually something to be salvaged from the ashes of their efforts – even if it is occasionally just a good belly laugh.

就是历史往往是由一群如醉如痴、勤勤恳恳的男男女女所推动的，他们永无止境地思考，由衷地认为自己的所想可以推动科学进步，为人类造福，为世界造福。正如本书所讲，他们并不总是正确的，但他们失败的故事常常是壮观的、充满永恒的魅力。他们的努力即使以失败告终，但后人也往往可以从中汲取养分，哪怕有时只是把它们当作一个很好的笑料罢了。

Golden Island

The Scots settlement in
AMERICA called New
CALEDONIA.
A.D. 1699 Lat. ... North

The Outward Bay

of Caledonia

Hoot Loch Cove

New
Edinburgh

The Inward Bay
of Caledonia

D A R I E N

English Miles

THE GR.

BAY

PART ONE

第一部分

~~CANCELLED~~
PIGEON-GUIDED MISSILES

注释

① Alps [ælps] n. 阿尔卑斯山
② conquer ['kɒŋkə(r)] vt. 征服，克服
③ Dickin Medal 迪金勋章
④ bacterial [bæk'tɪərɪəl] adj. 细菌的
⑤ bold [bəʊld] adj. 勇敢的，无畏的，boldest 为 bold 的最高级形式
⑥ renowned [rɪ'naʊnd] adj. 有名的，享有声誉的

Since Hannibal crossed the Alps① with his elephants in the third century bc, aiming to conquer② Rome, mankind has frequently used all sorts of animals as tools of war.

One creature to make a considerable contribution to the sphere of human conflict is the humble pigeon. During the Second World War some 250,000 homing pigeons served with British forces. Thirty-two were even awarded the Dickin Medal③, the military's version of the VC for animals. The US had its own Pigeon Service and one of its ranks, nicknamed GI Joe, was credited with saving 1,000 lives.

Yet some believed the pigeon's military capabilities lay beyond carrying messages. In 1945 an official at Britain's Air Ministry Pigeon Section, Lea Rayner, spoke of how pigeons might carry explosives and even become vehicles for bacterial④ warfare. But it was in America that the boldest⑤ designs for the use of pigeons against the enemy were formulated – by the renowned⑥ behavioural scientist B.F. Skinner. Through his work with the birds, Skinner believed that they held the key to perfecting the next step in military technology, the guided missile.

鸽子制导导弹

早在公元前3世纪，汉尼拔便率领大象群越过阿尔卑斯山，意图一举征服罗马。从此之后，人类便频频将各种动物用作战争工具。

有一种动物在人类战争领域作出了突出贡献，那就是不起眼的鸽子。在第二次世界大战期间，约有25万只信鸽服役于英国军队，其中32只甚至荣获堪称"动物部队维多利亚十字勋章"的**迪金勋章**。美国也有自己的战鸽队，其中一只绰号为"美国大兵"（G.I.Joe）的战鸽，战功赫赫，曾经拯救了上千人的性命。

然而，有些人认为鸽子的军事才能并不仅仅限于传递信息。1945年，英国航空部信鸽局官员利·雷纳谈到了鸽子可以携带炸药，甚至可以成为细菌战的运载体。在美国，著名的行为心理学家伯尔赫斯·弗雷德里克·斯金纳制订了用鸽子对付敌人的大胆计划。通过研究鸟类，斯金纳认为下一步改进军事技术（即制导导弹）的关键应在鸟类身上。

注释

迪金勋章，又名迪肯勋章，是授予动物界的最高军事奖章。

Burrhus Frederic Skinner, born in 1904, was no quack; and for a time the US authorities didn't think his ideas were bird brained either. They took him very seriously indeed and funded his pigeon project, hoping that harnessing① the abilities of these winged wonders would give them the critical edge in defeating the Axis powers, and the key to their country's future defence.

By the time the Second World War broke out, Skinner, who later went on to become a professor at Harvard, was already a successful psychologist. He had helped pioneer a theory called operant conditioning②. In essence he believed that animal and human behaviour was reinforced by external factors like rewards or punishments. Legend has it that, after seeing a flock of birds flying alongside a train, Skinner suddenly realised that his research on conditioning had a practical application for the war effort. He could train birds to guide missiles to their targets. 'It was no longer merely an experimental analysis. It had given rise to a technology,' he said.

In 1940 his initial③ work at the University of Minnesota had shown that pigeons could be trained to repeatedly pick out a target by using the reinforcement technique of pecking④ at pieces of grain. In this way they could be conditioned to keep pecking at a target and their movements linked to mechanisms that would alter the guidance controls of the missiles. One of Skinner's birds pecked at an image more than 10,000 times in forty-five minutes and, by 1941, he was able to show that it was possible to train pigeons to steer⑤ towards small model ships. Despite this, after taking his research to defence officials, he was told that his proposal 'did not warrant⑥ further development at this time'.

Then, in 1942, Skinner's work was suddenly dusted off by researchers looking for a psychologist to train dogs to steer anti-submarine torpedoes⑦. It eventually led to a grant of $25,000

斯金纳出生于1904年，绝对不是夸夸其谈之辈，而美国政府有关部门也一度认为他的想法言之有理。事实上，他们对这一设想倍加重视，还出资让斯金纳开展鸽子研究项目，希望开发鸽子能力可以赋予他们打败轴心国的决定性优势，可以成为美国未来国防的关键。

第二次世界大战爆发时，斯金纳已经是一位成功的心理学家了，后来又成为哈佛大学教授。在此之前，他创立了**操作性条件反射**理论。事实上，他认为动物和人类的行为均会受到奖励或惩罚等外部因素的强化。据说斯金纳在看到一群鸟沿着一列火车飞行之后，突然意识到他对条件反射的研究可能在战争中派上用场：他可以训练鸟类进行导弹制导。他说："这门理论不再停留在实验分析阶段，它已经催生了一项技术。"

1940年，他在明尼苏达大学的前期研究表明，通过采用啄食谷粒强化训练方法，能使鸽子反复选择同一目标。通过这种方法，可以在鸽子身上建立起反复啄咬同一目标的条件反射，再将其运动与可以改变导弹导航控制的机械装置相连。斯金纳的一只鸽子曾在45分钟内点啄一幅图像多达1万余次。1941年，他证明了可以训练鸽子向小轮船模型飞过去的可能性。尽管如此，当他把研究成果拿到国防部官员那里时，他们告诉他这项提议"此时

注释

操作性条件反射，亦称工具性条件反射，是由美国行为主义心理学家斯金纳于20世纪30年代在经典条件反射的基础上创立的实验方法。

▲ 行为心理学家斯金纳于20世纪40年代率先研发的鸽子制导导弹样品的鼻锥。

The nose cone of a prototype pigeon-guided missile, as pioneered by behavioural psychologist B.F.Skinner in the 1940s.

from the US Government to develop the idea via a company called General Mills①, who also wanted to do their bit for the war effort. During this further research, Skinner found that a pigeon would track an object by pecking at a screen even under all sorts of difficult conditions, including rapid descent② and the noise of explosions. Subsequently, plans were drawn up to experiment using the pigeons inside the new, aptly③ named, 'pelican④' missiles.

Skinner's plan was to load three pigeons into their own pressurised⑤ chambers⑥ inside the missile nose cone⑦. Lenses⑧ in the missile threw up an image of the target on a glass screen. Once they saw it the pigeons would start to peck and their movements translated into adjustments in the missile's guidance rudders⑨. The results amazed fellow scientists with some saying that the pigeons' accuracy rivalled⑩ that achieved by radar. In addition, using pigeons in this capacity didn't involve radio signals that could be jammed by the enemy.

Skinner himself explained why three birds were needed:

When a missile is falling toward two ships at sea, for example, there is no guarantee that all three pigeons will steer toward the same ship. But at least two must agree, and the third can then be punished for his minority opinion. Under proper contingencies⑪ of reinforcement a punished bird will shift immediately to the majority view. When all three are working on one ship, any defection is immediately punished and corrected.

注释

① General Mills 通用磨坊，是一家世界财富500强企业，主要从事食品生产业务
② descent [dɪ'sent] *n.* 下降
③ aptly ['æptlɪ] *adv.* 巧妙地
④ pelican ['pelɪkən] *n.* 鹈鹕
⑤ pressurise [p'reʃəraɪz] *v.* 对……施加压力，给……增压
⑥ chamber ['tʃeɪmbə(r)] *n.* 室
⑦ nose cone *n.* （火箭、飞机等的）前锥体，鼻锥体，头锥
⑧ lenses [lenz] *n.* 透镜，镜头（lens 的名词复数）
⑨ rudder ['rʌdə(r)] *n.* 船舵
⑩ rival ['raɪvl] *vt.&vi.* 与……竞争，竞争
⑪ contingency [kən'tɪndʒənsi] *n.* 可能性，偶然性（复数：contingencies）

不足以进行进一步研究。"

1942年，当研究人员需要一位心理学家训练狗引导反潜鱼雷时，斯金纳的研究又突然重见天日了。最终，美国政府拨款25000美元，由一家名为通用磨坊（General Mills）的公司研究这个想法，而这家公司也想为战争出点力。在进一步研究中，斯金纳发现鸽子即使身处快速下降、爆炸响声等恶劣条件下，也能通过点啄屏幕来跟踪目标。随后，他又制订了一项计划，把鸽子放在一种名副其实的新型"鹈鹕"导弹中进行实验。

斯金纳的计划是把三只鸽子放入导弹鼻锥内各自的增压室里，导弹中的镜头把目标图像投射到玻璃屏幕上。一看到图像，鸽子就开始点啄它，而它们的运动则用于调节导弹制导舵。实验结果让其他科学家大为惊奇，有人甚至认为鸽子的精确度绝不逊于雷达。此外，用鸽子制导无须使用可能受到敌方干扰的无线电信号。

至于为什么要用三只鸽子，斯金纳是这样解释的：

> 比如当一枚导弹朝海上的两艘船飞去时，我们无法确保三只鸽子都转向同一艘船，但至少两只鸽子的方向肯定是一致的，这样第三只鸽子就可能因为持少数意见而受到惩罚。在适当强化条件下，受罚的鸽子就迅速转向多数意见。当三只鸽子转向同一艘船上时，所有脱群行为都会立即受到处罚并得到纠正。

之后，导弹样品制造出来了，鸽子部队也准备就绪，30天之内就可以大规模生产这种导弹了。但是，在1944年的一次高层展示中，政府官员无论如何也无法理解鸽子为什么会如此听话。斯金纳说："让一只活生生的鸽子来执行任务，无论结局多么完

Prototype① missiles were built and legions② of pigeons prepared to be trained up for service. Mass production of the missiles was primed③ to snap④ into action in just thirty days. But, at a high level demonstration in 1944, government officials simply couldn't get their heads round the idea that pigeons could ever be satisfactorily controlled. Skinner said: "The spectacle⑤ of a living pigeon carrying out its assignment, no matter how beautifully, simply reminded the committee of how utterly fantastic our proposal was."

That wasn't quite the end of the project. In 1948 it was revived by the US Navy under the name Project Orcon (organic control). This time a pigeon's beak was fitted with a gold electrode⑥ which would hit a semi-conductive plate to report the target's position to the missile's controlling mechanism. But a line was finally drawn under the project in 1953 when further developments in electronic guidance systems made the pigeons redundant⑦.

Skinner's pigeon-guided missile system never took off. A plan for bat bombs, also tested during the 1940s, met a similar fate. The creatures were to be fitted with mini parachutes and timed incendiary⑧ devices and released en masse from aircraft. The idea was that, after being dropped over enemy cities, they would naturally look for places to roost⑨ in buildings and set them alight⑩, causing a firestorm. The $2 million project was abandoned not long after a colony of bats accidentally blew up a fuel tank at the Carlsbad Air Force base in New Mexico. Meanwhile, Soviet forces trained dogs to blow up German tanks. This produced mixed

注释

① prototype ['prəʊtətaɪp] *n.* 雏形
② legion ['li:dʒən] *n.* 众多、大批、大量
③ prime [praɪm] *vt.* 使准备好
④ snap [snæp] *vi.* 敏捷地动作，迅速地行动
⑤ spectacle ['spektəkl] *n.* 景象
⑥ electrode [ɪ'lektrəʊd] *n.* 电极，电焊条
⑦ redundant [rɪ'dʌndənt] *adj.* 多余的，累赘的
⑧ incendiary[ɪn'sendɪəri] *adj.* 引火的
⑨ roost [ru:st] *vi.* 栖息
⑩ alight [ə'laɪt] *vi.* 下来，飞落

美，只会让评审委员会觉得我们的提议是多么不切实际。"

这个项目并未就此偃旗息鼓。1948年，美国海军以"Orcon（有机控制）项目"重新启动了这个项目。这次在鸽子喙上安装了一个金电极，用其撞击一块半导体板，从而向导弹控制机构报告目标位置。但到1953年，随着电子制导系统的进一步发展，鸽子已显得十分多余，这个项目最终被画上了句号。

斯金纳的鸽子制导导弹系统从未得以实现，而同样在20世纪40年代开展的一项蝙蝠炸弹计划也遭遇了相似的命运。研究人员在蝙蝠身上装上微型降落伞和定时放火装置，然后从飞机上一齐放出。原本设想从敌方城市上空放下蝙蝠后，蝙蝠会像往常一样寻找建筑物栖息，于是点燃建筑物引发火灾。然而，由于一群蝙蝠意外引爆了新墨西哥州卡尔斯巴德空军基地的一个油罐，这项投资200万美元的计划不久就流产了。与此同时，苏军则训练狗炸毁德军坦克。结果令人喜忧参半：一些狗竟然掉头跑向己方阵地，因为它们训练时听到的是苏军坦克的引擎声，而不是德军坦克的引擎声。

不过在接下来的几十年里，世界各国军队从未放弃在战争中使用动物的努力。较近的一个例子是，苏联在太平洋沿岸的一个秘密军事基地对所谓的"海豚敢死队"进行训练，看它们是否能携带水雷攻击敌方军舰。这个项目随着苏联的解体而终止，敢死队中的一些海豚甚至被卖到了伊朗，但是人们至今都搞不懂伊朗人用这些海豚来做什么。

俄国在训练海豹查找水雷方面取得了较大的成功，而美国今天则使用海狮保护一个重要的海军基地免受恐怖分子袭击。这些受过训练的海狮，能将一个类似手铐的装置套在敌方潜水员身上，然后把他拖到附近的船上。服兵役的动物在日常生活中也会

results when some of the dogs ran towards their own lines: they'd been trained to listen for the engine noises of Soviet tanks, not enemy German ones.

The world's military did not give up on using animals in warfare in the ensuing[1] decades, however. More recently the Soviets experimented on training so-called kamikaze[2] dolphins at a secret base on the Pacific coast, to see if they could carry mines to attack enemy warships. The programme was halted[3] by the break up of the Soviet Union and some of the kamikaze dolphins were even sold to the Iranians, though quite what they did with them remains unclear.

Russians have trained seals to locate mines with more success, while the US today uses sea lions to protect one of its important naval[4] bases against terrorists. The sea lions have been trained to place a cuff[5], like a handcuff, on enemy divers, who can then be reeled[6] to nearby boats. Animals in military service also still save lives on a daily basis; in 2010 a black Labrador called Treo was awarded the Dickin Medal for his work sniffing[7] out bombs in Afghanistan.

Well after the Second World War MI5 kept a stock of trained pigeons for its security work. And of his experiments with pigeon-based warfare, B.F. Skinner himself admitted: 'I knew that in the eyes of the world we were crazy.' But, in an article in American Psychologist in 1960, he also told how he believed that many genuine scientific advances would never have been achieved without the odd 'crackpot[8]' idea. His research on pigeon-guided missiles certainly demonstrated how powerful the idea of conditioning could be. Incredibly, six years after the first pigeon project had ended, Skinner found that some of the pigeons he had trained could still identify the same targets.

注释

① ensue [ɪn'sjuː] vi. 接着发生，接踵而来
② kamikaze [ˌkæmɪˈkɑːzi] n. <日> 第二次世界大战期间空军敢死队
③ halt [hɔːlt] vt. 使停止，使中断
④ naval ['neɪvl] adj. 海军的
⑤ cuff [kʌf] n. <俚>手铐
⑥ reel [riːl] v. 卷，绕
⑦ sniff [snɪf] v. 嗅，闻
⑧ crackpot ['krækpɒt] adj. 古怪的，离奇的

拯救人们的性命。2010年，一只名叫特瑞欧的黑色拉布拉多猎犬因在阿富汗嗅探炸弹的功劳而被授予迪金勋章。

第二次世界大战之后，军情五处还保留了一些受过训练的鸽子，用于国家安全工作。而对于鸽子战实验，斯金纳自己也承认：“我知道在世人眼里，我们简直是疯了。”然而，在1960年《美国心理学家》的一篇文章中，他写道：“如果没有那些疯狂怪异的想法，许多真正的科学进步也许永远无法实现。”他在鸽子制导导弹方面的研究确实展现了条件反射这一思想的力量。令人难以置信的是，在首个鸽子计划结束六年后，斯金纳发现他训练过的一些鸽子仍能辨认出同样的目标。

ABANDONED
A SOUND PLAN FOR DEFENCE

注释

① bungalow ['bʌŋɡələʊ] *n.* 平房，小屋
② monolith ['mɒnəlɪθ] *n.* 独块巨石，整料
③ loom [luːm] *v.* 朦胧出现
④ concave [kɒn'keɪv] *adj.* 凹的
⑤ edifice ['edɪfɪs] *n.* （巨大而雄伟的）建筑物，大厦
⑥ eerie ['ɪəri] *adj.* 怪异的，神秘的
⑦ hush [hʌʃ] *n.* 安静，寂静
⑧ marshy ['mɑːʃiː] *adj.* 沼泽般的，湿软的
⑨ menace ['menəs] *n.* 威胁；恐吓
⑩ Luftwaffe ['luftvɑfə] *n.* <德>（纳粹时代的）德国空军
⑪ realm [relm] *n.* 领域，范围
⑫ acoustic [ə'kuːstɪk] *adj.* 声学的
⑬ carnage ['kɑːnɪdʒ] *n.* 大屠杀，残杀
⑭ trench [trentʃ] *n.* 沟，渠，战壕
⑮ rage [reɪdʒ] *vi.* （战争等）激烈进行
⑯ artillery [ɑː'tɪləri] *n.* 炮，炮兵部队

Just beyond a row of unremarkable looking bungalows①, near the lonely foreland of Dungeness on Britain's south coast, a group of strange concrete monoliths② loom③ out of a flat gravel landscape. Today these crumbling concave④ edifices⑤ provide an eerie⑥ spectacle in the hush⑦ of their marshy⑧ surroundings. They are, in fact, the bizarre remains of a cutting-edge defence system designed to protect the country from the roar of war.

Before radar famously helped Britain fend off the menace⑨ posed by the Luftwaffe⑩ in 1940, government technicians had been working on another top-secret technology to protect the country from airborne attack. All that is now left as a reminder of this fascinating chapter in the defence of the realm⑪ are a few imposing ruins. These huge acoustic⑫ mirrors, abandoned in the Kent countryside, were once the key to the plan.

The strange story of the sound mirrors begins amid the carnage⑬ of the trenches⑭. In 1915, as the First World War raged⑮, a bright solider with a degree in physics called William Sansome Tucker was working with a unit experimenting on sound ranging – a way to detect the location of enemy artillery⑯ from the sound waves produced by shell fire. Tucker managed

声波防御计划

在英国南部海岸邓杰内斯寂静的海滩上，坐落着一排不起眼的平房，在这排平房的前方，矗立着一群奇怪的混凝土庞然大物。今天，这些破碎的凹形巨大建筑在周围一片潮湿而寂静的环境中呈现出一种怪异的景象。事实上，这些奇异残骸是一种为了使英国免遭战争破坏而设计的尖端防御系统。

众所周知，1940年，雷达使英国避开了德国空军的威胁，但在此之前，为使国家免受空袭破坏，政府技术人员已经在着手研究另一项绝密技术，而现在只留下了这些壮观的废墟，让人们回想起那段保家卫国的华丽篇章。这些巨大的声反射镜，曾经是防御计划的关键，现在则被遗弃在肯特郡的荒野之中。

声反射镜的离奇故事始于战壕里的士兵大量战死期间。1915年，第一次世界大战战火正炽，一位具有物理学学位的聪明士兵威廉·桑萨姆·塔克（William Sansome Tucker）正在一个小分队试验声波测距——通过轰炸产生

注释

① pinpoint ['pɪnpɔɪnt] vt. 准确地指出，精准定位

② retaliatory [rɪ'tælɪətrɪ] adj. 报复的，以牙还牙的

③ enhance[ɪn'hɑːns] vt. 提高，增加，加强

④ extraneous[ɪk'streɪnɪəs] adj. 外部的，外来的

⑤ harness['hɑːnɪs] vt. 利用，给（马等）套轭具，控制

⑥ stethoscope ['steθəskəʊp] n. 听诊器

⑦ compartment [kəm'pɑːtmənt] n. 隔间（尤指火车车厢中的）

⑧ horizontally [ˌhɒrɪ'zɒntəlɪ] adv. 水平地，横地

⑨ vertically ['vɜːtɪklɪ] adv. 垂直地，直立地

⑩ squadron ['skwɒdrən] n. 一对，一群

to come up with a 'hot wire' microphone which eventually allowed units to pinpoint① artillery to about 50m, allowing more accurate retaliatory② attacks. Meanwhile, back in Britain, experiments with sound mirrors were already afoot. The Zeppelin threat to Britain had led a Professor Mather to start looking into huge dishes which would focus the sound of enemy aircraft and airships so that they could be heard, though still well out to sea, before they came into visual range. In 1916 he'd had a crude 16ft mirror dug into a chalk hillside in Kent, and went on to do early work with concrete versions of these sound mirrors. One of these early designs was able to pick up a plane 10 miles away.

After the war, Tucker got the job of Director of Acoustical Research at Air Defense Experimental Establishment, where he was able to bring together his work on microphones and the sound enhancing③ capability of the satellite-dish-style reflectors. His efforts were centered on a site near Dungeness. This location, just yards from the coast, was felt to be far enough away from extraneous④ noise, and in 1928 Tucker oversaw the construction of a concrete mirror, 20ft in diameter. The general idea was that the low frequency sound waves from aircraft would be harnessed⑤ and concentrated, then picked up by a microphone. A man would stand next to the mirror listening in with a stethoscope⑥ and, in this way, be able to give the alert when aircraft were approaching. A bigger, 30ft mirror was built in 1930. Its larger surface area made it more accurate, and the 'listener' was able to work from inside a sound-proofed compartment⑦. By moving the microphone collector horizontally⑧ and vertically⑨ using foot pedals and a wheel, he could identify the direction the sound was coming from and get a bearing on, say, a squadron⑩ of incoming aircraft.

In the same year another huge mirror with a different style

的声波探测敌军炮兵位置。塔克研制出了一种"热线"麦克风，最终能将敌军炮兵的位置精确到50米，以便更加精确地回击。同时，声反射镜的实验也在英国紧锣密鼓地进行。由于英国受到齐柏林硬式飞艇的威胁，有一位叫马瑟（Mather）的教授开始尝试用巨大的圆盘聚集敌军飞机、飞艇的声音，这样就能提前听到海面上视野范围之外的飞机、飞艇的声音。1916年，他在肯特郡白垩岩的山腰处建了一面4.88米（16英尺）口径的声反射镜，并进一步研究混凝土版的声反射镜。在这些早期的设计中，有一面声反射镜可以侦测到16.09千米（10英里）以外的飞机。

战后，塔克担任防空实验研究所声学研究室主任。在那里他得以把麦克风与卫星圆盘式反射镜增强声音的功能结合起来。他的研究以邓杰内斯附近的一个地点为中心。这个离海岸只有几米远的地点足以远离外部噪声。1928年，塔克监督建造了一面6.10米（20英尺）口径的混凝土声反射镜。总体思路是：探测并集中飞机的低频声波，然后传到麦克风。人站在声反射镜前用听诊器监听，这样在飞机接近时就能起到预警作用。1930年，塔克建造了一面更大的，9.14米（30英尺）口径的声反射镜。表面积增大使其定位更加精确，监听者也可以在隔音室中工作。通过用脚踏板和轮子横向或纵向移动麦克风，收音器就能确定声音来自哪个方向，并确定敌机的方位。

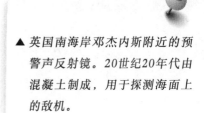

▲ 英国南海岸邓杰内斯附近的预警声反射镜。20世纪20年代由混凝土制成，用于探测海面上的敌机。

Acoustic early-warning mirrors near Dungeness on Britain's south coast. Constructed from concrete in the 1920s, they were designed to detect enemy aircraft out to sea.

注释

① finesse [fɪˈnes] v. 巧妙地做
② hamper [ˈhæmpə(r)] v. 妨碍，阻止
③ interfere [ˌɪntəˈfɪə(r)] vi. 干预，干涉，妨碍
④ echoing [ˈekəʊɪŋ] n. 回声（波）现象
⑤ salvo [ˈsælvəʊ] n. 齐射的炮火
⑥ prompt [prɒmpt] v. 促使

of design was built – 200ft long and 26ft high with a curve of 150°. In this design a set of twenty microphones were built in to help pick up the sounds. The mirror's new design also enabled it to pick up sounds with the longer wavelengths usually created by aircraft. In 1932 this mirror was able, on one occasion, to locate aircraft 20 miles away. The human ear could only pick them up around 6 miles away. Tucker was beginning to finesse① the technology.

For all the improvements in his designs, however, the mirrors were still hampered② by the fact that sound waves travel relatively slowly. By the time the aircraft had been plotted it was already very close. In the years that Tucker had been working on the project aircraft speed had also been gradually increasing, making the sound mirrors less effective. There might, for instance, be only around four minutes' warning before the identified aircraft were overhead. Even at the lonely spot he'd chosen for the experiments the noise of wind, ships at sea and even car traffic could interfere③ with the results too.

Tucker nevertheless carried on. Perhaps the words of MP Stanley Baldwin that 'the bomber will always get through' were echoing④ in his ears. He recommended a line of mirrors should be built along the coast from The Wash to Dorset, and, in 1934, the government made preparations to construct the first salvo⑤ of these defensive mirrors around the Thames Estuary. There was to be a string of 200ft mirrors at 16 mile intervals with smaller 30ft-style mirrors filling the gaps.

In 1935 came news that was to make Tucker's work look like an expensive white elephant. While Tucker had been slaving away on his sound mirrors, scientist Robert Watson-Watt had been developing radar. His work had been prompted⑥ by the air ministry who wanted to know about the chances of developing a death ray which could take out enemy aircraft.

同年，另一面风格不同的巨大声反射镜建成，长60.96米（200英尺），高7.92米（26英尺），弯曲度为150度。这次的设计内置了20个麦克风，用于收集声音。新的设计也能够探测到更长的波长，这种波长通常是飞机产生的。1932年，这面声反射镜曾有一次定位到32.19千米（20英里）以外的飞机。人耳最远只能听到9.66千米（6英里）远的声音。因此，塔克开始完善这项技术。

然而，无论如何改进他的设计，声反射镜始终受到声波传递速度相对较慢的影响。等飞机位置确定时，它已经离得非常近了。在塔克研究这项技术的几年中，飞机的飞行速度也逐渐加快，这使得声反射镜效力更不明显，也许只能在探测到的飞机到达头顶上空之前4分钟预警。即使在他选择试验的偏远地点，风、海上船只甚至汽车发出的噪声也会影响效果。

然而塔克继续坚持他的研究。或许是因为国会议员斯坦利·鲍德温（Stanley Baldwin）的话——"轰炸机总是有效的"一直在他耳畔回响。他建议应该沿着从瓦士湾（沃什湾）到多塞特的海岸建一列声反射镜。1934年，政府准备初次环绕泰晤士河口建一列防御声反射镜。每隔25.75千米（16英里）建一个60.96米（200英尺）的大口径声反射镜，中间再建一些9.14米（30英尺）的小口径声反射镜。

1935年，一则消息竟然使塔克的研究成了累赘。正当塔克为他的声反射镜埋头苦干时，物理学家罗伯特·沃森·瓦特（Robert Watson-Watt）研制出了雷达。航空部试图研制一种死亡射线来压制敌机，所以才有了沃森·瓦

注释

① intercept [ˌɪntəˈsept] vt. 拦截，拦住
② beam [biːm] n. 光线，束，柱
③ bounce [baʊns] vi. 跳，反弹
④ detect [dɪˈtekt] vt. 查明，侦查出
⑤ redundant [rɪˈdʌndənt] adj. 多余的，累赘的
⑥ Acoustical Research Station 声学研究站
⑦ hastily [ˈheɪstɪlɪ] adv. 匆忙地，仓促地
⑧ thwart [θwɔːt] vt. 阻挠，使受挫折，挫败
⑨ invasion [ɪnˈveɪʒn] n. 入侵，侵略
⑩ allied [ˈælaɪd] adj. 同盟的，同盟国的

Instead of knocking aircraft out of the sky, Watson-Watt decided to see if radio waves could be used to find out where they were so that fighters could intercept① them. His system, which built on the theories of others, involved sending out beams② of radio waves, then monitoring when they bounced③ off approaching objects. At first his system detected④ aircraft at 8 miles, then 40 miles, and eventually 200 miles. Almost overnight the sound mirrors had become redundant⑤, and in 1939 the Acoustical Research Station⑥ was closed. The authorities even ordered that Tucker's remaining concrete mirrors be blown up. Thankfully, for military historians, these orders were ignored.

During the Battle of Britain, in the summer of 1940, the string of radar stations, which had been hastily⑦ built on the back of Watson-Watt's work, helped thwart⑧ Hitler's planned invasion⑨. Using radar, enemy raids could be quickly pinpointed and intercepted by fighter aircraft. It also helped that the head of the Luftwaffe, Hermann Goering, decided to switch his strategy away from attacking radar stations to targeting the nation's cities instead. While destructive for Britain, these raids allowed more successful attacks on German bomber squadrons. Sir William Sholto Douglas, who went on to lead Fighter Command during World War Two, said: 'I think we can say that the Battle of Britain might never have been won … if it were not for the radar chain.'

Tucker's work had not been entirely in vain, however. During military exercises involving the sound mirrors in the early 1930s he had helped develop an early warning network to relay information on enemy activity to a central command. This concept, which was turned into a highly effective network with radar, was one of the key reasons why radar became such a useful tool for Allied⑩ forces in the Second World War and, arguably, helped turn the tide of the conflict.

特的研究。但沃森·瓦特没有研究如何把飞机打下来，而是决定研究一下无线电波是否能用于确定敌机位置，这样就可以拦截它们。他的系统以其他人的理论为基础，通过发出无线电波束，可以监控电波遇到接近的物体时反射回来的电波。开始时，他的系统能探测到约13千米（8英里）外的飞机，接着是约64千米（40英里），最终达到322千米（200英里）。几乎一夜之间，声反射镜就变得毫无用处。1939年，政府关闭了声学研究站，甚至命令炸掉塔克的混凝土声反射镜。多亏了军事历史学家，这条命令最终没有执行。

1940年夏，在英德战争期间，紧跟着沃森·瓦特的研究建成了一系列雷达站，帮助挫败了希特勒的入侵计划。战斗机利用雷达可以快速定位敌机并予以拦截。这也使得德国空军的头目赫尔曼·戈林（Hermann Goering）决定把战略目标从袭击雷达站转为袭击城市。虽然这对英国来说是毁灭性的，但这些空中雷达探测系统使得对德国成群结队轰炸机的打击更为成功。第二次世界大战时仍然领导战斗机司令部的威廉·谢尔托·道格拉斯（William Sholto Douglas）爵士说："我想，如果没有雷达链，我们可能永远打不赢不列颠之战。"

但是，塔克的研究也并非完全徒劳。20世纪30年代初使用声反射镜的军事演习期间，他研制出一种预警系统，将敌军活动的信息传送到中央指挥部。这个概念后来发展成为高效的雷达网络，这也使雷达在第二次世界大战中成为盟军的重要工具，成功扭转了战争的局势。

REJECTED

THE DIABOLICAL DEATH RAY

In 1916 an American named Albert Bacon Pratt was granted a patent for a new, hair-raising invention – the helmet① gun. The firearm② was, said Pratt, 'adapted to be mounted on and fired from the head of the marksman'. The user would be able to shoot instantly by tugging③ on a wire attached to his mouth. Usefully, continued Pratt, the device could also double up as a cooking pan. The idea was never heard of again – perhaps because the recoil④ of the gun would surely have broken the neck of anyone foolish enough to use it. Almost as bizarre was another US patent, filed in 1862, for a combined plough and gun. The basic idea was that you could get on with tilling⑤ the earth safe in the knowledge that you were fully prepared if marauding⑥ soldiers suddenly attacked across your fields.

It seems that mankind is always aiming to invent ever more fiendish⑦ ways of waging war. One 'Holy Grail' of warfare, however, has been the attempt to develop a so-called death ray. Greek inventor Archimedes supposedly used a 'burning mirror' death ray against Roman ships during the Siege of Syracuse in 214–212 B.C. If he did do so, it can't have worked very well because the Romans ultimately triumphed in that contest. And

残忍的死亡射线

1916年，美国人阿尔伯特·培根·普拉特（Albert Bacon Pratt）获得了头盔枪的专利，这是一项令人毛骨悚然的新发明。普拉特说："这种枪适用于把发射装置安装在射手头顶，然后从这里进行射击。"使用者只要拉动连到嘴里的电线就能立刻进行射击。普拉特还说，这个设备还可以当锅使用，非常实用。不过再也没听说过这个主意——可能因为假如有人笨到胆敢使用它的话，枪的后坐力肯定会折断他的脖子。美国另外一项怪异的专利是1862年注册的犁枪组合。其基本思路是：如果兵匪突然打到你的地里，有犁枪一体的装备做保障，你就可以放心地继续耕你的地了。

人类似乎一直在尝试发明更加残忍的作战工具，然而战争的"圣杯"之一却是试图发明所谓的死亡射线。传说希腊发明家阿基米德（Archimedes）在公元前214—公元前212年的叙拉古保卫战中，就用过"燃烧镜"反射的死亡射线来对抗罗马战船。如果事实真是这样，那么效果肯

注释

① scramble ['skræmbl] vi. 攀登，爬
② Gloucestershire ['glɒstəʃiə] n. 格洛斯特郡（英国英格兰郡名）
③ tussle ['tʌsl] n. <口>扭打，争斗
④ chancellor ['tʃɑ:nsələ(r)] n. 大臣
⑤ exchequer [ɪks'tʃekə(r)] n. （英国）财政部
⑥ doom [dʊm] v. 注定
⑦ tireless ['taɪələs] adj. 不疲倦的
⑧ polymath ['pɒlimæθ] n. 博学者
⑨ deterrent [dɪ'terənt] n. 制止力量，威慑力量
⑩ zap [zæp] n. 精力，活力

it wasn't until the late nineteenth century and early twentieth century that the idea of an all-powerful death ray really caught the imagination, not only of science fiction writers like H.G. Wells, but also some of the brightest inventors of the day.

One of those who led the scramble① to develop a workable death ray was the Gloucestershire② born inventor Harry Grindell Matthews. He is a now half-forgotten figure. Yet in the early 1920s, his tussles③ with the British military establishment over the device became a newspaper editor's dream – a story that just ran and ran. Matthews, born in 1880 and married twice, once to an opera star, was certainly an eccentric figure. But he was also very clever. Initially an electronics engineer, he gained fame after inventing a wireless communication machine called the aerophone – a device which Matthews said was able to send messages to planes from the ground. Matthews even proudly showed it off to the then chancellor④ of the exchequer⑤ and future prime minister David Lloyd George. Matthews did early work on talking movies too, recording an interview with Ernest Shackleton in 1921 just before the explorer set off on a doomed⑥ final trip to Antarctica. A tireless⑦ polymath⑧, his luminaphone was a machine that used beams of light to make music.

Matthews also knew a thing or two about battlefields, having served in the Boer War. In fact it was his determination, after being injured himself on the battlefield, that mankind should avoid more wars. This fuelled his efforts to create a weapon that, using a powerful electrical beam, could disable aircraft engines in mid-flight. If mastered it would provide each nation with the ultimate deterrent⑨.

During the First World War the British government offered £25,000 to anyone who could come up with machines to remotely control aircraft or zap⑩ attackers in the sky, from the

定不好，因为最终获胜的是罗马人。直到19世纪末和20世纪初，这种全能死亡射线的想法才不仅勾起了H.G.威尔斯（H.G.Wells）等科幻小说家的无限遐想，也激起了当时最聪明的发明家的奇思妙想。

在众多争相发明切实可行的死亡射线的先行者中，有一位出生于英国格洛斯特郡的发明家，名叫哈里·格林德尔·马修斯（Harry Grindell Matthews）。现在人们几乎已经完全遗忘他了，然而在20世纪20年代，报社编辑做梦都想报道他和英国军事部门围绕一套设备的争执——这件事情被人们争相传颂。马修斯生于1880年，结过两次婚，有一次娶的是位歌剧明星。他绝对是个怪人，但他非常聪明。他刚开始是电子工程师，因为发明了一种名为空中电话的无线通信设备而名声大噪——马修斯说这种设备可以从地面向飞机发送信息。他甚至自豪地向时任财政大臣、后来任首相的戴维·劳合·乔治（David Lloyd George）炫耀这一设备。马修斯还对有声电影做出了贡献，他录下了对探险家欧内斯特·沙克尔顿（Ernest Shackleton）1921年最后一次有去无回的南极探险出发前的采访。他是一位孜孜不倦的博学者，他发明的光声机是一种能够利用光束产生音乐的机器。

马修斯曾经参加过布尔战争，因此对战场也略知一二。事实上，他在战场受伤后就坚决地认为人类应该避免更多的战争。这激发他努力研制出一种武器，用强大的电子束使飞行中的飞机发动机无法工作。如果掌控得当，这将对各个国家都产生终极威慑。

第一次世界大战期间，英国政府悬赏2.5万英镑，希

注释

① herald ['herəld] v. 传达，预示……的到来

② raid [reɪd] n. 急袭，突然袭击

③ admiralty ['ædmərəltɪ] n.（英国）海军部海军上将职位

④ pop [pɒp] vi.（意外地、突然地）出现

⑤ allegedly [ə'ledʒɪdlɪ] adv. 据说，依其申诉

⑥ forthcoming [ˌfɔːθ'kʌmɪŋ] adj. 即将到来的

⑦ divulge [daɪ'vʌldʒ] vt. 泄露，暴露

⑧ ionize ['aɪəˌnaɪz] v.（使）电离

⑨ soak up 吸收

⑩ twist [twɪst] vt. 扭成一束，搓，捻，绕，卷

⑪ backer ['bækə(r)] n. 支持者，赞助者

⑫ bulb [bʌlb] n. 电灯泡

⑬ rumour ['ruːmə(r)] n. 传闻，谣言

⑭ con [kɒn] adj. 欺诈的，骗取信任的

ground. They were particularly worried about the new threat from the air heralded① by German Zeppelins, which killed over 550 people during raids② over the country's major cities between 1914 and 1918. Matthews seized his chance to enter the fray. He developed a wireless boat which impressed admiralty③ officials when he demonstrated it on a London lake in 1915. Despite paying out the £25,000 prize, however, the government never developed the craft any further.

Later, in 1923, Matthews popped④ up again with a new invention to tackle the airborne threat as military aircraft were developing fast. He suddenly announced his idea of a death ray to a select group of journalists. They reported that in a demonstration his creation had successfully stopped a motorbike engine at a distance of 50ft. Matthews announced: 'I am confident that if I have facilities for developing it I can stop aeroplanes in flight.' In February 1924 The Air Ministry asked him to exhibit his death ray to their experts. Matthews refused. This may have dated back to his annoyance at allegedly⑤ finding a military official tampering with his famed aerophone back at another demonstration in 1911. Crucially, though, he wasn't forthcoming⑥ in divulging⑦ the science behind his death ray. In the ensuing media frenzy there was talk of ionized⑧ air and short radio waves, but Matthews kept up the mystery on how his beam worked while soaking up⑨ the limelight.

After some arm twisting⑩ by his backers⑪, Matthews was finally persuaded to reveal his death ray in the spring of 1924. It was not the dramatic spectacle the ministry scientists had been expecting. In the lab his beam did appear to switch on a light bulb⑫ and cut off a small motor, but government officials weren't impressed. Some said the science wasn't new and, with rumours⑬ of con⑭ tricks doing the rounds, the government went cold on the idea. Matthews decided enough was enough. He

望能制造出一种从地面遥控飞机或在空中袭击入侵者的武器。他们尤其担忧德国齐柏林飞艇带来的新威胁，这种飞艇在1914—1918年对英国主要城市的空袭中致使550多人死亡。马修斯抓住这次机会，研制出一艘无线遥控船，1915年在伦敦某湖上的演示给海军军官留下深刻印象。虽然英国政府支付了2.5万英镑赏金，但未对该艇作进一步研发。

1923年，随着军用飞机迅速发展，马修斯突然又有了新发明，能对付空中的威胁。他突然向一群选定的记者宣布了死亡射线的想法。报道称，在演示时马修斯的新发明成功地使15.24米（50英尺）之外一辆摩托车的引擎停止工作。马修斯宣称："如果再有一定设施改进这项发明，我有信心让飞机停止飞行。"1924年2月，航空部请他向部里的专家展示死亡射线，但马修斯拒绝了。这可能是由于在1911年的一次展示中，他怀疑一位军官擅自动了他著名的空中电话。不过关键是他还不打算公布死亡射线背后的科学原理。在之后媒体的不断报道中，有人谈到电离空气和无线电短波，但马修斯始终对光束运行原理守口如瓶，吊着众人的胃口。

1924年春天，在资助人的不断施压下，马修斯最终同意公开死亡射线。不过那次展示并没有像航空部的科学家所期待的那么激动人心。在实验室里，死亡射线确实可以使灯泡发光，也可以使一台小型发动机停止工作，但并没有给政府官员留下深刻的印象。有人说这种技术并不是创新，甚至传言说这是一场骗局，政府对这个创意也变得冷淡了。马修斯决定适可而止，把这项发明卖给法国。他的

注释

① plead [pli:d] *vi.* 恳求

② saga ['sɑ:gə] *n.* 萨迦（中世纪冰岛和挪威关于历史事件、历史人物、轶事传闻等的北欧传说），英雄传奇

③ melodramatic [ˌmelədrə'mætɪk] *adj.* 戏剧性的

④ retort [rɪ'tɔ:t] *vt.* 反驳，报复

⑤ rodent ['rəʊdnt] *n.* 啮齿目动物（如鼠、松鼠、河狸等）

⑥ flick [flɪk] *n.* 轻弹，轻拂

⑦ festive ['festɪv] *adj.* 节日的，过节似的，喜庆的

⑧ gimmick ['gɪmɪk] *n.* 花招，诡计

⑨ flop [flɒp] *n.* <口> 彻底的失败

⑩ recluse [rɪ'klu:s] *n.* 隐居者，遁世者

would sell his invention to the French. His team of financial backers pleaded① with him not to, even racing to the airport to try and stop him.

Meanwhile the press were having a field day with the saga②. The *Daily Express* screamed of 'Melodramatic③ Death Ray Episodes' on its front page. Questions were being asked in the House of Commons too – shouldn't the government be making more effort to keep the death ray British? The government was forced to back down. It offered Matthews £1,000 if he could show that his ray would stop a motorcycle petrol engine. Matthews, now in France, retorted④ that he was doing a deal with the French. Indeed his associate there, Eugene Royer, filed a patent for a death ray around this time. In addition, that summer Matthews made a film with Pathé News showing off the shiny death ray machine and how it could kill a rodent⑤ at the flick⑥ of a switch.

Yet, strangely, a finished weapon never materialised, either in France or back in Britain. Later the same year, and still famous, Matthews set off for a fundraising tour of the US. Again, however, he refused to explain or show how his death ray worked. Over the next few years, interest in his claim and the offers of cash gradually faded away. Then, on Christmas Eve 1930, Matthews was back in the headlines. Above the skies of London festive⑦ shoppers could make out the words 'Happy Christmas' projected onto the clouds. It was the work of Matthews' new project – the sky projector. This time his beam wouldn't be killing things: it would help sell them. It seemed like a great advertising gimmick⑧. What is more, there couldn't be any question that it worked. Sadly for Matthews it proved a commercial flop⑨ and he was declared bankrupt in 1934. When Matthews died in 1941 he had become a virtual recluse⑩, locked away in a remote compound in Wales and still beavering away

资助人请求他不要这样做，甚至追到机场去阻止他。

与此同时，新闻界大肆宣传起这件事来。《每日快报》在头版刊登了醒目的标题"戏剧性的死亡射线"。下议院也有人发难——政府难道不应该努力让死亡射线留在英国吗？英国政府被迫作出让步，表示如果马修斯能展示死亡射线使摩托车的汽油发动机停止工作，就给他1000英镑。马修斯当时正在法国，强硬地回应说他正在和法国做交易。事实上，他的法国合伙人欧仁·鲁瓦耶（Eugene Royer）当时确实为死亡射线申请了专利。此外，1924年夏天，马修斯和百代电影公司合作了一部电影，炫耀那部很炫的死亡射线机器如何一合开关就能把老鼠杀死。

然而奇怪的是，无论是在法国还是在英国，自始至终也未做出一件成品的武器。1924年年底，依然颇负盛名的马修斯动身前往美国筹资，但他仍然拒绝展示或透露死亡射线的工作原理。接下来的几年里，对他的言论感兴趣并愿投资的人越来越少。1930年圣诞夜前夕，马修斯又出现在报纸头条。节日购物的人可以在伦敦上空看到投影到云上的"圣诞快乐"字样。这是马修斯新项目的杰作——天空投影仪。这次，他的射线不再用于杀死东西，而是用于促销东西，这似乎是个不错的广告噱头，而且一定会管用的。然而马修斯也真够倒霉的，这个项目竟然也失败了，他于1934年宣告破产。1941年去世时，他已经成为一名真正的隐士，把自己锁在威尔士一处偏远的院子里，仍然在孜孜不倦地想着新的计划。

继马修斯之后，还有许多人倡导死亡射线。1924年，美国人埃德温·R.斯科特（Edwin R. Scott）声称他是

on new schemes.

Many more exponents of death rays had followed Matthews. In 1924 an American, Edwin R. Scott, claimed that he had actually been the first with a 'lightning device' that could 'bring down planes at a distance'. In the early '30s the Spanish-born scientist Antonio Longoria reckoned[1] his death ray could kill birds 4 miles away. There were many other outlandish[2] claims. In the end, none of these colourful characters were able to convincingly[3] show that a death ray could work.

Despite this, even the renowned scientist Nikola Tesla, whose earthquake machine is the subject of the next chapter, believed it could be done. In 1934 he theorised that a concentrated beam of tiny particles formed of mercury[4] or tungsten[5] could be accelerated by a high-voltage[6] current and used to knock whole squadrons of planes out of the air, hundreds of miles away. This teleforce, as he called it, would 'afford absolute protection' to countries from any form of attack. But Tesla, already in his late 70s, was to die before he or anyone else could turn his concept into a reality. Just a few years later, nuclear weapons would overtake the mania[7] around death rays and, once again, they were consigned[8] to the realm of fiction and fantasy.

注释

① reckon ['rekən] *vt.* 认为

② outlandish [aʊt'lændɪʃ] *adj.* 古怪的，奇异的

③ convincingly [kən'vɪnsɪŋlɪ] *adv.* 令人信服地，有说服力地

④ mercury ['mɜːkjəri] *n.* 汞，水银

⑤ tungsten ['tʌŋstən] *n.* 钨

⑥ high-voltage [haɪ 'vəʊltɪdʒ] *adj.* 高压的

⑦ mania ['meɪnɪə] *n.* 狂热

⑧ consign [kən'saɪn] *vt.* 用作，当作

真正发明可以"远程击落飞机"的"闪电装置"的第一人。20世纪30年代初，西班牙科学家安东尼奥·朗格利亚（Antonio Longoria）认为他的死亡射线能杀死6.44千米（4英里）外的鸟。除此之外，还有很多其他古怪的断言，但最后没有一个人能令人信服地展示死亡射线可以发挥作用。

尽管如此，就连著名科学家尼古拉·特斯拉（Nikola Tesla）也相信可以发明出死亡射线，他发明的地震机是另一篇的主题。1934年，他从理论上宣称，高压电流可以将汞或钨组成的微粒子集束加速，用于打击数百英里之外的飞行中队。他把这种武器称为"远程突击队"，可以为各国提供针对各种形式攻击的"绝对防护"。然而直到**特斯拉**年近八旬去世之时，他本人或其他人也未能将这一概念变成现实。仅仅几年之后，核武器就抢了死亡射线的风头。死亡射线再次成为科幻小说中的话题。

编者注

特斯拉（1856—1943）享年86岁。此处依原文译。

~~FAILED~~
THE MISPLACED MAGINOT LINE

注释

① André Maginot 安德烈·马其诺
（1877—1932），法国政治家
② veteran ['vetərən] n. 老兵
③ bombard [bɒm'bɑːd] vt. 炮击，
轰炸
④ foe [fəʊ] n. 敌人
⑤ vulnerable ['vʌlnərəbl] adj.（地
方）易受攻击的
⑥ fortification [ˌfɔːtɪfɪ'keɪʃn] n. 防
御工事
⑦ deride [dɪ'raɪd] vt. 嘲笑，愚弄
⑧ parliament ['pɑːləmənt] n. 议
会，国会
⑨ ominously ['ɒmɪnəslɪ] adv. 恶兆
地，不吉利地，预示地

In the 1920s André Maginot① was a worried man. A veteran② of the First World War trenches, he'd seen his home town of Revigny-sur-Ornain in the region of Lorraine, near the border with Germany, bombarded③. He had also been badly wounded in the leg at the bloody battle of Verdun on the Western Front. Maginot was determined that the 'rape' of France at the hands of its old foe④ would never happen again. He felt that the Treaty of Versailles – later blamed for leaving Germany in a bitter frame of mind that fuelled Hitler's rise to power – didn't go far enough, and that France was still vulnerable⑤ to attack from a Germany that simply couldn't be trusted. Maginot would dedicate much of the rest of his career to championing the idea of a massive line of fortifications⑥ along the border between the two countries. Fortunately for him, Maginot never lived to see it become one of history's most derided⑦ military concepts.

Maginot became a member of the French parliament⑧ when he returned to civilian life after the First World War and, by the late 1920s, was Minister of War. The idea of a defensive line had been around for a while. Ominously⑨, Philippe Pétain, then Inspector General of the army, was one of its fans. As

错位的马其诺防线

20世纪20年代，安德烈·马其诺（André Maginot）整日忧心忡忡。他是在第一次世界大战的战壕里摸爬滚打过的老兵，曾经目睹位于法德边界洛林的家乡奥尔南河畔的勒维尼（Revigny-sur-Ornain）遭到轰炸，在西部前线凡尔登（Verdun）的浴血战斗中腿部受了重伤。马其诺下定决心，宿敌侵略法国的历史绝不能重演。他认为，强迫与德国签订的《凡尔赛条约》（后来被指责为使德国心生怨恨，激起希特勒掌权的原因）远远不够，而且德国根本不可信，仍然可能攻击法国。马其诺把他剩余的职业生

注释

《凡尔赛条约》，全称为《协约国和参战各国对德和约》，是第一次世界大战后战胜国（协约国）对战败国（同盟国）的和约。条约于1919年6月28日在巴黎的凡尔赛宫签订，于1920年1月10日正式生效。

marshal[1], Pétain would later become infamous for leading the subjugated[2] government which collaborated with the Germans following France's surrender in 1940. But a fortified line had its critics too. Most of the French communists and socialists were against it. Others, like Charles De Gaulle, would warn that it left France cowering[3] behind its defences; he favoured a more mobile army. Maginot warned that French manpower in the 1930s would be no match for Germany's bigger population if conflict came, and that a defensive line would keep the peace by deterring an invasion. It would create jobs too and, if it had to be defended, could be done by smaller numbers of men rather than the massive armies which had been needed to repel[4] the Germans in 1914–1918.

In 1926, funding was given for some experimental sections of line and in December 1929 Maginot outlined his plan for a full-scale project:

> We could hardly dream of building a kind of Great Wall of France, which would in any case be far too costly. Instead we have foreseen powerful but flexible means of organizing defence, based on the dual[5] principle of taking full advantage of the terrain[6] and establishing a continuous line of fire everywhere.

Maginot's arguments convinced enough French members of parliament and by 1930 he'd swung[7] a vote to give 3 billion francs to roll out a massive line of fortifications along the country's eastern border.

When Maginot died on 7 January 1932, reportedly from complications caused by eating bad oysters[8], work was in full swing. During the decade of the Maginot Line's construction, four million truckloads of earth and concrete would be moved.

注释

① marshal ['mɑːʃl] n. 元帅
② subjugate ['sʌbdʒugeɪt] vt. 征服，降伏
③ cower ['kaʊə(r)] vi. 畏缩，抖缩
④ repel [rɪ'pel] vt. 击退
⑤ dual ['djuːəl] adj. 双的，两部分的
⑥ terrain [tə'reɪn] n. 地形，地势
⑦ swung [swʌŋ] v.（使）摇摆（swing 的过去式和过去分词）
⑧ oyster ['ɔɪstə(r)] n. 牡蛎

涯都奉献在倡导沿法德边境修筑大规模防线上。幸运的
是，他没能活着看到这道防线成为历史上最可笑的军事
思想之一。

第一次世界大战结束后，马其诺退伍，成为法国国
会成员。20世纪20年代，他任国防部部长。他考虑修建
防线的想法有段日子了。当时的陆军总长菲利普·贝当
（Philippe Pétain）是这个想法的拥护者之一，这不是个
好兆头（当时的元帅贝当后来因1940年任总理时，向德国
投降而声名狼藉）。不过也有人批评防线，多数法国共产
党员和社会党党员都反对修筑防线。其他反对者，例如夏
尔·戴高乐（Charles De Gaulle）警告说，这样会使法国
躲在防线后面当缩头乌龟，他赞成提高部队的机动能力。
马其诺警告说，20世纪30年代假如打起仗来，法国的参
战人数根本无法与人口众多的德国抗衡，而防线可以打消
德国的入侵念头，从而维持和平。此外，这样还可以增加
就业岗位，发生战事时只用少数人就可以防御，而不用像
1914—1918年那样需要大批部队才能击退德国。

1926年，国会为几段实验性防线的修建提供了资金；
1929年12月，马其诺简要介绍了整个工程计划：

> 我们几乎无法想象修建一座法国的长城，这
> 无论怎样都太过昂贵。相反，我们已经预见到有
> 力而灵活的防御手段，既能充分利用地形，又能
> 在任何地方建立连续的战线。

注释

① casemate ['keɪsmeɪt] *n.* 炮台，炮塔

② retract [rɪ'trækt] *vt.& vi.* 撤回或撤销

③ trap [træp] *n.* 圈套

④ ouvrages [uvra:ʒ] *n.* 工事

⑤ garrison ['gærɪsn] *n.* 守备部队，守卫

⑥ dignitary ['dɪgnɪtəri] *n.* 高官，显要人物

⑦ chunk [tʃʌŋk] *n.* 相当大的数量或部分

⑧ impenetrable [ɪm'penɪtrəbl] *adj.* 不能通过的

⑨ masterstroke ['mɑːstəstrəʊk] *n.* 妙举，巧妙的动作

⑩ outflank [ˌaʊt'flæŋk] *v.* 翼侧包围

The eventual fortifications, extending for hundreds of miles from Switzerland to Belgium, as well as along the Italian border, were impressive. Built of concrete and steel, the Line was not a continuous structure but involved a total of some 500 separate buildings: including a string of huge forts as well as artillery casemates①, retracting② machine gun turrets and tank traps③, all close enough to be able to support each other. The main forts, known as ouvrages④, each had a garrison⑤ of up to 1,000 men and often extended to several storeys underground. The fortifications housed everything from hospitals to living quarters connected by 100 miles of tunnels, even featuring underground railways. There were other fortified 'lines' like the German's Siegfried Line, later know as the 'Westwall'. But France's new defences were unmatched anywhere in the world and they proudly showed them off to visiting dignitaries⑥.

In the end it was strategic errors that would ruin the Maginot Line's reputation, not the failings of these fortifications themselves. Belgium and France were allies, so the fortifications were light along their shared border leaving a large chunk⑦ of Northern France unprotected, should the Germans invade that way. The Belgians had their own defences and the forest of the Ardennes, which covered the area just to the north of the Maginot Line, was so impenetrable⑧ (so the thinking went) that the Germans surely wouldn't be able to pull off an invasion that way. If the Germans did come through Belgium then France's mobile forces could be concentrated in the area to see them off.

When Hitler finally turned his attention to the west in May 1940, his generals were, indeed, respectful enough not to try and attack the Maginot Line head on. Instead they came up with a strategic masterstroke⑨, outflanking⑩ both the Line and the French and British armies distracted in northern Belgium. In a surprise strike they quickly captured Belgium's key Fort Eben

马其诺的论述说服了足够多的法国国会成员。到1930年，他扭转了投票，成功给这项工程申请到30亿法郎，用于沿法国东部边境修建大规模防御工事。

马其诺于1932年1月7日离世，据悉是死于吃了坏牡蛎引起的并发症。当时马其诺防线的修建工程正全面推进，10年内要移走400万辆货车的土和混凝土。这条防线从瑞士到比利时，还沿着意大利边境，绵延数百千米，蔚为壮观。防线由钢筋混凝土建成，并非是一个连续结构，而是包括大约500座独立建筑：包括一系列巨型碉堡和炮塔、隐蔽机枪射塔和反坦克堑壕，相互间距很近，可以互相支援。那些主要碉堡被称作"工事"，每一座工事的守卫都达1000人，地下一般都延伸了好几层。整座防御工事包括医院、宿舍等各种设施，应有尽有，通过约160千米（100英里）的地下通道连接，甚至还有地下铁轨。世界上还有其他防线，例如德国的齐格菲防线，后被称为"西墙"。但法国的新防线是全世界任何防线都无法比拟的，他们自豪地向来访政要展示这条防线。

最终令马其诺防线声名扫地的是战略错误，而不是这些防御工事本身。由于当时比利时和法国是盟国，所以沿着两国边境而建的防御工事就比较薄弱，假如德国从这里入侵，法国北部的一大片地区未受到防护，将处于危险之中。比利时有自己的防御工事，**阿登高地**森林覆盖的地区正好位于马其诺防线的北部，据说它是不可穿越的天然屏障，德国肯定无法从那里入侵。如果德国真的穿过比利时而来，法国的机动部队就可以在那个区域集结，击退敌人。

注释

阿登高地是位于法国东北部、比利时东南部及卢森堡北部的森林台地，面积约1万平方千米。

Emael with an attack by paratroopers[1]. Hundreds of Panzer tanks then swept through Ardennes forest, making a mockery[2] of the myth that it was impassable. Within days the Germans were bearing down on Paris. Hitler's forces could now attack the Maginot Line to the south from the rear[3] and, while many of its forts held out, by June France had surrendered. Tens of thousands of French soldiers were taken prisoner from the Maginot defences without having had the chance to fire a shot.

Like many great military blunders[4] it's easy, with the benefit of hindsight, to criticise the Maginot Line. Debate rages[5] among historians about whether it was ill-conceived and whether it worked. It certainly failed to save France. Historian Ian Ousby, for instance, has called it a 'dangerous distraction of time and money when it was built, and a pitiful irrelevance when the German invasion did come in 1940'.

Others have argued that the fortifications were so impressive that it gave rise to an atmosphere of complacency[6] in France, a so-called 'Maginot mentality', so that it didn't plan properly for war or build a sufficient mechanised mobile force to back up the Line.

It's a common observation among military analysts that nations are 'always planning for the last war'. The Maginot Line certainly seems to fall into this category, devised by men who had lived through the largely static First World War in which 1.4 million of their countrymen had perished[7]. They had not planned for the highly mobile nature of the Second World War with its blitzkriegs[8]. Other arguments suggest that the Maginot Line wasn't the problem, it was the French army which hadn't been organised well enough, or that the air force hadn't been strong enough to support the fortifications.

注释

① paratrooper ['pærətruːpə(r)] n. 伞兵
② mockery ['mɒkəri] n. 嘲笑，愚弄
③ rear [rɪə(r)] n. 后部，背面，背后
④ blunder ['blʌndə(r)] n. 失策，疏忽
⑤ rage [reɪdʒ] vi. 流行，风行
⑥ complacency [kəm'pleɪsnsi] n. 自满，满足
⑦ perish ['perɪʃ] vi. 死亡，丧生，毁灭
⑧ blitzkrieg ['blɪtskriːg] n. 闪电战，闪击战

1940年5月，希特勒最终把注意力转向西部，他的将领们挺有礼貌，没有正面进攻马其诺防线，而是出奇制胜，绕过防线两端，在比利时北部吸引英法部队。德国空降兵突然袭击，很快占领了比利时的埃本-埃马尔要塞。数百辆德国坦克横扫阿登森林，对其不可穿越的神话来说是一个莫大的讽刺。德军几天之内就进占巴黎。希特勒的部队这时就可以从背后攻击马其诺防线南段了，虽然防线内有很多碉堡在抵抗，但到6月法国就投降了。成千上万的法国士兵还没来得及开枪就从马其诺防线的防御工事里被抓出来，投进监狱。

像很多其他严重的军事错误一样，对马其诺防线做马后炮式的批评是容易的。历史学家激烈地争论这条防线是否设想就有问题，是否发挥了作用。它当然没能拯救法国。例如历史学家伊恩·乌斯比（Ian Ousby）曾说它"修建的时候费时伤财，而1940年德国真正入侵时又毫无用处。"

还有人说是因为防御工事太壮观，法国处在自豪的氛围中，有所谓的"马其诺心态"，所以没有很好地为战争做准备或者组建有效的机械化部队以支援防线。

军事分析家经常发现，各国"总是在为上次战争做准备"。马其诺防线似乎也落入这一窠臼，防线设计者从基本静态的第一次世界大战中死里逃生，而140万同胞死亡。他们完全没有为以闪电战为特征的第二次世界大战做好准备。也有人说不是马其诺防线本身的问题，而是法国部队组织不利或空军不够强大，无法支援防御工事。无论真相如何，从某种意义上来说，马其诺修建了一座"法国

注释

① rebel ['rebl] *n.* 叛乱者，造反者

Whatever the truth, in one way, Maginot had constructed a 'Great Wall of France'. Despite what he had said, his Line shared something in common with the Great Wall of China. That great structure, for all its strength, didn't save the country from invasion either. In 1644 the Ming dynasty was overthrown when a rebel① general merely opened a pair of its gates, allowing an invading army through to overrun the country.

的长城"。不管马其诺说了什么，这道防线和中国的长城
有一些共同点，那座伟大的建筑最终也没能保卫住国家。
1644年，一个叛军将领只是打开一扇城门，放敌军入侵中
原，就推翻了明朝。

~~CANCELLED~~
THE GREAT 'PANJANDRUM'

注释

① formidable ['fɔ:mɪdəbl] *adj.* 可怕的，令人敬畏的
② contraption [kən'træpʃn] *n.* 奇妙的装置，新发明
③ smash [smæʃ] *vt.* 打碎，撞击

The department of wheezers and dodgers had a problem. The Nazis' Atlantic Wall – a series of fortresses and batteries, bunkers and mines, running down Europe's west coast from Scandinavia to northern Spain – must be breached if Britain was to make headway in the Second World War. For the team charged with creating a weapon to breach it, this was a formidable① task. Happily, the wheezers and dodgers (more formally known as the Directorate of Miscellaneous Weapons Development, DMWD) had an exemplary track record in creating contraptions② of carnage. The Hajile, which dropped heavy equipment from the skies using rockets, and the Hedgehog, an anti-submarine bomb, were their inventions. And though the Atlantic Wall, which included concrete walls 10ft high and 7ft thick in places, would take some smashing③ at its strategic points, the DMWD had the talent, the willpower and the authority to meet the challenge.

Not least, they could call on the services of C.R. Finch-Noyes; full-time wing commander, part-time inventor, who had already developed 'hydroplane skimmers'. These short-range, high-capacity torpedoes were designed to obliterate Germany's

武器中的"大亨"

插科打诨局（正式名称为杂项武器开发管理局，DMWD）遇到了难题。如果英国想要在第二次世界大战中有所进展就必须攻破纳粹德国的"大西洋墙"，这是一条欧洲西海岸上从斯堪的纳维亚到西班牙北部的防线，由一连串的碉堡、火炮阵地、沙坑和地雷组成。对于负责研发攻破这一防线的武器的团队来说，这是一项非常艰巨的任务。幸好插科打诨局在发明武器方面颇有经验，他们曾经发明了采用火箭空投重型设备的武器哈吉勒和名为"刺猬"的反潜艇炸弹。尽管"大西洋墙"的一些战略要点非常坚固，一些部位混凝土墙甚至高3米（10英尺）、厚2米（7英尺），但插科打诨局仍然有能力、有毅力、有权力来迎接这次挑战。

更重要的是，他们可以向英国皇家空军中校C.R.芬奇-诺伊斯（C.R. Finch-Noyes）求助。他是一位业余发明家，曾研发了水上飞机掠行艇。这种短程高性能鱼雷可以快速穿过水下，摧毁德军的默内河谷水坝和埃德尔

注释

① propel [prə'pel] vt. 推进，推动
② forerunner ['fɔ:rʌnə(r)] n. 先驱，先驱者
③ buster ['bʌstə(r)] n. 庞然大物，破坏者
④ warhead ['wɔ:hed] n.（尤指导弹的）弹头
⑤ fourfold ['fɔ:fəʊld] adj. 四倍的
⑥ sketch [sketʃ] v. 简短地陈述或描写
⑦ verse [vɜ:s] n. 诗
⑧ smithereen [ˌsmɪðə'ri:n] vt. 把……击碎，把……炸成碎片
⑨ mastermind ['mɑ:stəmaɪnd] vt. 策划，谋划
⑩ cart [kɑ:t] n. 运货马车

Möhne and Eder dams by propelling① themselves violently across water before reaching an exquisitely destructive end. While considerably less famous than the 'bouncing bomb' later developed by Barnes Wallis (and the subject of a renowned British war film), Finch-Noyes' missiles were the forerunner② of Wallis' dam buster③ bomb. By developing such missiles, Finch-Noyes had learned how much more magnified explosions are when they occur underwater. He also believed in packing as big a punch as possible and so had increased the explosives in the warheads④ of torpedoes fourfold⑤.

Puzzling over the Atlantic Wall challenge, he sketched⑥ an impression of the kind of weapon that would be required to hit thick, high concrete with sufficient force to breach an area large enough to get troops and equipment through. Most problematic of all was how to propel the bomb from landing craft at sea to the foot of the wall on the beach in the face of German gunners. His sketches turned into what became known as the 'Panjandrum', a reference to nonsense verse⑦ written almost 200 years earlier by actor Samuel Foote. Packed with explosives, Finch-Noyes' Panjandrum would nip out of the ocean from a landing craft and, once at the foot of the wall, blast all around to smithereens⑧.

Masterminding⑨ the whole project was Sub-Lieutenant Nevil Shute Norway – who went on to write the novel *A Town Called Alice* under the name Nevil Shute – and he liked Finch-Noyes' design. An explosive wooden cart⑩, with 10ft-tall wheels, powered by rockets and able to travel at 60 mph, hadn't been attempted before. This was the kind of impressive machine Shute needed: big enough to carry more than 1,000 kilograms of explosives and deliver its deadly load more or less in the right direction. Testing would have to be thorough and on terrain similar to the beaches of Normandy, where the Panjandrum

河谷水坝。虽然芬奇-诺伊斯发明的导弹不如巴恩斯·沃利斯（Barnes Wallis）后来研发的跳弹（这也是一部著名英国战争电影的主题）出名，但这是沃利斯研发的大坝毁灭炸弹的前身。芬奇-诺伊斯研发这种导弹时，研究了水下发生爆炸的威力的扩大程度。他还认为发射力度要尽可能大，因此将鱼雷弹头内的炸药增加了三倍。

　　他苦思"大西洋墙"的挑战，大致描述出需要这样一种武器：要能够攻击又高又厚的混凝土，威力大到打开的缺口能让部队和装备通过。最大的问题是怎样在德军枪手的眼皮底下把炸弹从海上的登陆艇发射到海滩上的墙根处。他描述的武器成了众人所知的"大亨"，这个词引自约200年前演员萨缪尔·福特（Samuel Foote）写的打油诗。芬奇-诺伊斯的"大亨"装上炸药就能从登陆艇上冲出海面，一到墙根就把周围炸得粉碎。

　　精心策划这项工程的是内维尔·舒特·诺威（Nevil Shute Norway）中尉——他后来以内维尔·舒特署名写了小说《爱丽丝城》——他喜欢芬奇-诺伊斯的设计。以前从没人尝试过能爆炸的木头马车，装有3米（10英尺）高的车轮，用火箭驱动，行驶速度可以达到97千米/时（60英里/时）。这正是舒特需要的那种惊人武器：大到可以装下1000多千克炸药，基本能发射到指定方位。此外，还要周密安排试验，由于计划把"大亨"首先部署在诺曼底海滩，所以试验地点的地形要与诺曼底的地形相似。芬奇-诺伊斯的部队——联合作战试验研究所（或简称COXE，这个名称很适合当时的情形）驻扎在德文郡的阿普尔多尔，正好知道有一个叫"嗬，一路向西"

注释

① dune [dju:n] n. 沙丘

② compromise ['kɒmprəmaɪz] vi. 折中解决，妥协，退让

③ convoy ['kɒnvɔɪ] n. 护航队

④ trundle ['trʌndl] vt. 使滚动，运送

⑤ pied piper 穿花衣的吹笛手，善开空头支票的领导者

⑥ chagrin ['ʃægrɪn] n. 懊丧，悔恨

⑦ extravagant [ɪk'strævəgənt] adj. 奢侈的

⑧ deploy [dɪ'plɔɪ] vt.& vi. 使展开（尤指军事行动），施展

⑨ erratically [ɪ'rætɪklɪ] adv. 不规律地，不定地

⑩ stutter ['stʌtə(r)] vt.& vi. 结结巴巴地说，不顺畅地工作

⑪ tenuous ['tenjuəs] adj. 薄的，细的，精细的，稀薄的

⑫ scratch [skrætʃ] v. 搔痒，抓，扒

⑬ velocity [və'lɒsəti] n. （沿某一方向的）速度

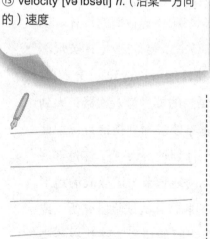

was scheduled to be deployed first. Finch-Noyes' unit, the Combined Operations Experimental Establishment – or COXE as it was appropriately called in the circumstances – based at Appledore, Devon, knew of such a beach: Westward Ho! A west-country holiday heaven, with swathes of beautiful golden sand, high surf, sand dunes①, a bit of shingle and some mud, it was everything that Normandy offered, and all on COXE's doorstep.

But secrecy, up to this point absolute, was compromised② once the prototype Panjandrum, constructed in East London, hit the road en-route to the resort: the sight of an oversized cotton reel with two giant wheels and explosives in the middle raising eyebrows as the convoy③ trundled④ south-west. Like the Pied Piper⑤, the Panjandrum attracted followers by the score, to the chagrin⑥ of the military. Ahead of testing, the DMWD implored onlookers to stay away for their own safety, to which the good people of Devon responded magnificently by turning up in droves to watch what many hoped to be the most extravagant⑦ firework display the town had ever seen.

Tests began modestly. Sand replaced explosives to start with, and low power rockets – just nineteen of them initially – were deployed⑧ for safety reasons. This was fortunate. On the first run, after careering erratically⑨ along the beach, the rockets fell off one wheel and the Panjandrum stuttered⑩ in every direction at the same time. On each subsequent test, control was tenuous⑪ to say the least. Nevil Shute Norway scratched⑫ his head and called for a pause for regrouping, which turned into a fundamental reassessment of the Panjandrum's design. What this weapon needed, he concluded, was another wheel.

Returning to the beach, the new and improved, three-wheeled, higher velocity⑬ Panjandrum carrying many more rockets, fared no better. The test team rolled the contraption

的海滩符合这样的要求。那是位于英国西部乡村的度假天堂，有美丽的金色沙滩、高高的海浪、沙丘、少量的鹅卵石和软软的淤泥，诺曼底所有的一切这里都有，并且就在COXE的门前。

截至目前，这个地方仍然是保密的，然而一旦在伦敦东部组装的"大亨"样品上路往这里来，这个度假胜地就不再是秘密：只见一个过大的线轴装了两只巨轮，中间填装炸药，让人惊讶万分。像穿花衣的吹笛手一样，"大亨"吸引了众多追随者，使军方懊恼。测试开始前，插科打诨局恳求旁观者为自身安全考虑离远点，德文郡的人反应很热烈，成群结队地前来旁观，很多人希望这会是从未见过的最奢侈的烟火。

测试小心翼翼地开始了。为了安全起见，刚开始用沙子代替炸药，用威力低的火箭，开始只用19枚。幸运的是在第一轮测试中，"大亨"沿海滩没方向地急速乱窜后，一只轮子上的火箭滑落了，朝四面八方突突地响。在后来的每次测试中，这种情况都很难控制。内维尔·舒特·诺威挠着头，要求重新组装，最后不得不从根本上重新考虑"大亨"的设计。他推断说，这武器需要再多装一个轮子。

经改良的新款"大亨"装了3个轮子，转速更高，携带更多火箭，回到海滩后，还是没有更好的表现。测试组成员朝低潮标志转动这个奇妙的装置，把周围清理干净，用手指堵住耳朵，然后开火。什么动静都没有。他们试了一次又一次，还是什么动静都没有。甚至潮水打到了脚踝，时间耗尽，还是什么动静都没有。最终，海

注释

① engulfing [en'gʌlfɪŋ] adj. 吞噬的

② deluge ['delju:dʒ] vt. 使淹没，使泛滥

③ brass [brɑːs] n. 要员，高层人员

④ detach [dɪ'tætʃ] vt. 分离，拆开

⑤ haywire ['heɪwaɪə(r)] n. 捆干草用的铁丝

⑥ bigwig ['bɪgwɪg] n. 有重大影响的人

⑦ binoculars [bɪ'nɒkjələ(r)z] n. 双筒望远镜

⑧ hurtle ['hɜːtl] vi. 猛冲，猛烈碰撞

⑨ dislodge [dɪs'lɒdʒ] vt. 把……逐出，驱逐

⑩ spun [spʌn] vi. 快速旋转，眩晕（spin的过去式及过去分词）

⑪ mown [məʊn] v. 刈，割（mow的过去分词）

⑫ ablaze [ə'bleɪz] adj. 着火的，发光的

towards the low-tide mark, cleared space around them, put their fingers in their ears, and fired. Nothing. They tried again and again. Still nothing. Even with the incoming tide lapping around their ankles and time running out, nothing. Finally, with the water all but engulfing[①] the Panjandrum, testing was abandoned and the machine left deluged[②] until the sea receded once more; the team looking on from the shoreline and wondering when the war would be over.

After three further weeks of modifications, the Panjandrum sported an improved steering system, seventy 20lb rockets, and agreement from top brass[③] that accuracy wasn't altogether a pre-requisite when blowing up German walls. This time Nevil Shute Norway was confident. Firing the Panjandrum from a landing craft towards the beach, the test began well, until several rockets once again detached[④] themselves and went haywire[⑤]. Some narrowly missed the heads of startled spectators; others exploded underwater. Norway scratched his head again and decided that the third wheel was probably unnecessary after all.

A series of later test runs highlighted new problems, some more serious than others, but the team eventually declared themselves satisfied. One more test was required in front of their superiors, and in January 1944 military bigwigs[⑥] left their Whitehall bunkers for the beach at Westward Ho! Focusing their binoculars[⑦] on the landing craft, they may well have been giving themselves a small congratulatory pat on the back when the test got off to a credible start. The Panjandrum hurtled[⑧] from the sea at wondrous speed, sparks flying from its rockets. But then old habits kicked in. As rockets dislodged[⑨], the Panjandrum spun[⑩], and shock replaced the confidence of those attending. Generals fled for cover. The official cameraman was almost mown[⑪] down. And the Panjandrum, rockets flailing, wheels ablaze[⑫], disintegrated.

水几乎吞没了"大亨",测试停止,任由机器留在那儿被海水淹没,直到海水再次退潮。小组成员在海岸线上望着,想着战争什么时候才能结束。

又用了三周时间改良"大亨",有了改进的操纵系统,可装载70枚9千克(20磅)重的火箭,并且军方高层同意炸德国防御墙时精确度不是必须要求的。这次内维尔·舒特·诺威有信心了。从登陆艇向海滩发射"大亨",测试开始顺利地进行,但后来几枚火箭又脱离了,开始乱飞。有些几乎擦着受惊的旁观者的头顶飞过,其他的在水下爆炸。诺威又挠头了,认为可能根本不需要那第三只轮子。

在后来的一系列测试中,"大亨"出现了明显的新问题,一个比一个严重,但测试小组最后宣布他们满意了。因为要在他们长官面前进行一次测试,1944年1月,军方要员们离开白厅(英政府所在地)的地堡,前往"嘀,一路向西"海滩。他们用望远镜注视着登陆艇,看到测试一开始比较靠谱,本来可以给自己庆祝式的赞许了。"大亨"从海上急速猛冲出去,火箭上火花飞溅,但随后又是老毛病——火箭脱离原位,"大亨"打转。要员们由自信变得震惊,将军们四散逃开寻找掩体,官方的照相师差点死了。"大亨"也碎裂了,火箭乱摆,轮子着火。

插科打诨局担忧恐慌,叫停了"大亨",似乎这就结束了。但2009年的某一天,又重新组装了"大亨"。在诺曼底登陆65周年纪念日,在"嘀,一路向西"海滩发射了重组的"大亨",这次稍小,而且幸好爆炸力也

Shell shocked, the department of wheezers and dodgers called time on the Panjandrum and that looked like the end for it until, in a one-day-only revival① in 2009, a Panjandrum was constructed once more. To mark the 65th anniversary of the D-Day landings, a slightly smaller and thankfully less explosive reconstruction of the calamitous② launch took place at Westward Ho! With the Union flag fluttering proudly on the sands, a small crowd gathered to see if it could repeat the exciting experience of its wartime launch. As the rockets powered up in a Catherine Wheel③ display of smoke and flames, the situation looked promising. But the drum hardly hurtled off its ramp④ in a manner likely to frighten passing Germans, propelling itself just 50m before shuddering⑤ to a halt. It did, however, travel in the right direction, no rockets chased after cameramen and no one feared for their lives.

注释

① revival [rɪ'vaɪvl] *n.* 复活，再生，再流行
② calamitous [kə'læmɪtəs] *adj.* 灾难的，悲惨的
③ Catherine Wheel 轮转烟火，烟花
④ ramp [ræmp] *n.* 土堤斜坡
⑤ shudder ['ʃʌdə(r)] *vt.& vi.* 战栗，发抖，震动

小一些。英国国旗在沙滩上自豪地飘扬，有一小群人聚集前来观看是否会出现战时发射时那样刺激的情景。随着火箭点燃，呈现出轮转式五彩烟火，情况看起来还不错。这圆桶以一种可能吓到经过的德国人的架势向前直冲，不过几乎还未冲出坡道，只推进了50米就开始振动然后静止。不过，它倒确实是朝着正确方向推进的，火箭也没有追着照相师跑，没人有性命之忧了。

DISASTER
THE DARIEN DEBACLE

In the late seventeenth century the Scots attempted to found a trading colony in Central America in an audacious① bid② to become a world power. Little did they know, this bold scheme would end in grizzly③ death and financial disaster. It was one of the factors which ultimately led to the Act of Union between Scotland and England in 1707, effectively bringing an end to hopes of Scottish independence for the next 300 years.

The Darien scheme was forged④ in the late 1690s by the Scotsman William Paterson, who had, ironically, been one of the founders of the Bank of England. Working in London around this time he'd met the explorer, buccaneer⑤ and surgeon Lionel Wafer who had just returned home after many years at sea. Wafer wrote a book called *A New Voyage and Description of the Isthmus⑥ of America*, describing his adventures in the region, and he inspired wealthy merchant Paterson with tales of an unconquered paradise ripe⑦ for the picking in the new world. According to Wafer the isthmus of Darien, in modern day Panama, was fertile, harboured friendly natives and was potentially the key to new trading routes to the spice-rich Far East. Here only some 50 miles separated the Atlantic and Pacific

一场灾难
达连计划的失败

　　17世纪末，苏格兰为了成为世界强国，曾冒险尝试在中美洲建立一个贸易殖民地。殊不知，这个大胆计划的最终结果是人员的大量死亡和经济危机，也成为最终导致1707年苏格兰和英格兰通过《联合法案》的一个因素，泯灭了之后300年苏格兰独立的希望。

　　达连方案形成于17世纪90年代末，发起者是苏格兰人威廉·帕特森（William Paterson）。具有讽刺意味的是，他曾是英格兰银行的创始人之一。在伦敦工作期间，他遇见了历经多年海上生活刚刚回国的探险家、海盗和外科医生莱昂内尔·韦弗（Lionel Wafer）。韦弗写了一本书，叫《美洲地峡的新航行与描述》，描述他在美洲的历险过程，而书中这个未被征服的新世界天堂故事激起了富商帕特森的强烈兴趣。根据韦弗所说，达连地峡（位于今天的巴拿马）土地肥沃，住着友好的土著人，而且可能是通往盛产香料的远东地区新的贸易途径。大西洋和太平洋在这里只隔约80千米（50英里）。如果能在这块狭长的土地上

注释

① monarch ['mɒnək] n. 君主，帝王
② famine ['fæmɪn] n. 饥荒，饥饿
③ lucrative ['lu:krətɪv] adj. 获利多的，赚钱的
④ sanction ['sæŋkʃn] vt. 批准，鼓励
⑤ publicist ['pʌblɪsɪst] n. 政论家，时事评论员
⑥ loftily ['lɒftɪlɪ] adv. 高地，崇高地
⑦ infectious [ɪn'fekʃəs] adj. 传染的，有传染性的
⑧ expedition [ˌekspə'dɪʃn] n. 考察，远征
⑨ embark [ɪm'bɑ:k] v. 上飞机，上船

coasts. Establishing a route across this narrow strip of land would make long trips round the stormy tip of South America needless. Those that controlled Darien might dominate trade across the two oceans.

At this time Scotland and England shared a monarch①, but Scotland had a separate parliament and could, technically, go its own way. But with famine② and years of costly civil war, the country was suffering economically. Meanwhile the English, Portuguese and Spanish had been busy snapping up lucrative③ new territories around the globe. Already a successful merchant, Paterson came up with a plan. He would raise money for a colony called New Caledonia, a venture to make the Scots proud and, most importantly, rich again. In June 1695 it was sanctioned④ by the Scottish parliament.

Paterson was a good publicist⑤ and his enthusiasm for the loftily⑥ titled 'Company of Scotland Trading to Africa and The Indias' was infectious⑦. While opposition from the rival East India Company saw off English and foreign investors, subscriptions from patriotic Scots worth £400,000 poured in to fund an expedition⑧ to Darien. People from even the lowliest levels of society signed up and many invested their life-savings. Estimates differ, but at least one fifth of Scotland's entire wealth was probably ploughed into the project.

In July 1698 Paterson was ready. Most of the 1,200 people who left with him didn't have any idea exactly where they were going until well into the journey. But, lured by the promise of 50 acres of land each, they embarked⑨ from Edinburgh in a fleet of five ships – the *St Andrew*, *Unicorn*, *Caledonia*, *Endeavour* and *Dolphin* – destined for a brave new world. The first aim was to start a settlement and get the colony on its feet with a view to establishing a new free trade port in Darien. In time, with trade busily passing through, the Scottish backers could then clean up

建立一条通道，人们就无须长途跋涉，绕过风急雨骤的南美洲最南端，来跨越这两个大洋了。谁能控制达连，谁就能支配这两个大洋之间的贸易。

这时，苏格兰和英格兰处于同一位君主的统治之下，不过苏格兰有自己的议会，因此从理论上说，苏格兰可以按自己的意愿行事。然而，由于饥荒和多年内战，苏格兰正处于经济危机之中。而此时，英国人、葡萄牙人和西班牙人一直在全球扩张殖民地，进行大肆掠夺，积累财富。成功的商人帕特森想出了一个计划。他想筹集资金建立一个叫新喀里多尼亚的殖民地，此举既能让苏格兰人自豪，更重要的是，还能让他们重新富强起来。1695年6月，这项计划得到了苏格兰议会的批准。

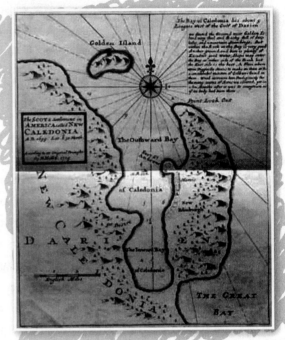

帕特森是一名出色的宣传家，他的非洲和印度苏格兰贸易公司听上去不同凡响，而他对公司的热情也充满了感染力。虽然来自竞争对手东印度公司的对抗赶跑了英格兰和外国投资者，但为资助去达连的一次远征，那些爱国的苏格兰人的捐款就已达40万英镑。连一些社会最底层的人都报名参与了这项投资，许多人还拿出了他们的毕生积蓄。这个项目到底投入了多少钱，至今尚无定论，但很有可能至少有全苏格兰1/5的财富。

1698年7月，帕特森一切准备就绪。在踏上旅程之前，随行的1200人中大部分并不知道他们要去哪儿。但

▲ 18世纪的一幅地图，上面绘有中美洲的达连地区，是苏格兰失败的殖民地。

An eighteenth-century map showing the territory of Darien in Central America, the site of Scotland's failed colony.

注释

① fort [fɔːt] n. 堡垒，要塞

② cannon ['kænən] n. 大炮

③ malaria [mə'leərɪə] n. 疟疾

④ wreak [riːk] v.（对某事物）造成破坏

⑤ havoc ['hævək] n. 大破坏，浩劫

⑥ mangrove ['mæŋɡrəʊv] n. 红树林

⑦ swamp [swɒmp] n. 沼泽

⑧ wig [wɪɡ] n. 假发

⑨ émigré ['emɪɡreɪ] n. 逃亡者，流亡者

with tidy commissions.

In November Paterson's group, minus seventy passengers who had died on the voyage, set foot on the shores of Darien. They lost no time building a fort① to protect themselves, armed with fifty cannons②. And a settlement, New Edinburgh, was founded. When news of their arrival reached Edinburgh on 25 March 1699, it was met with much celebration. The colonists began negotiating with the locals, and Paterson was among those who wrote back in positive terms about the state of the expedition that spring, perhaps not wanting to alarm the investors.

The optimism was not to last, however. The wet season soon arrived and tropical diseases like malaria③ wreaked④ havoc⑤ among the settlers; ten people were dying a day. Meanwhile, attempts to grow crops had proved tricky in a landscape thick with mangrove⑥ swamps⑦ and jungle. Settlers later recalled supplies riddled with maggots and one man told how his shoulders were covered in boils due to overwork. Attempts to trade for supplies with the local Kuna people also failed. Given the harsh climate it was not surprising that the natives were not impressed by the goods on offer: combs, mirrors, wigs⑧ and woollen clothes. The Scots soon resorted to hunting giant turtles to survive.

There was another problem. A mountain range stood in the way of any trading route that was to be pushed through to the Pacific, even if there had been enough settlers strong enough to build one. After eight months the decision was taken to abandon Darien. Of the 1,200 who had set out, only 300 of the original émigrés⑨ would survive to set foot in Scotland once again. Not knowing the fate of the first colony, a second fleet had left Scotland in August 1699 with another 1,302 settlers. It arrived in November to find Darien almost entirely deserted, bar a

是，公司许诺他们每人可以得到约20公顷（50英亩）土地，在这样的诱惑下，他们在爱丁堡登上了一支由5艘船（"圣安德鲁号""独角兽号""喀里多尼亚号""奋进号"和"海豚号"）组成的舰队，前往一个美丽的新世界。为了在达连建立一个新的自由贸易港，他们的首要目标是开创殖民地和使殖民地步入正轨。随着往来贸易不断繁荣起来，苏格兰投资者很快就能得到可观的报酬。

1698年11月，帕特森带领队伍（航行中已有70名乘客丧生）踏上达连的海岸。一到那里，他们立即修筑了一座碉堡自卫，碉堡里配备了50门大炮。接着，他们又建了一个名为"新爱丁堡"的殖民地。1699年3月25日，他们抵达达连的消息传到了爱丁堡，人们对此大肆庆祝。殖民者开始了与土著人的谈判。也许是不想惊动投资者，与许多人一样，帕特森在书信中只提及了那年春天他们探险过程中的好消息。

然而，这种乐观情绪没能持续多久。雨季很快到来，像疟疾之类的热带疾病在移民者中散布开来，造成了重大伤亡，平均每天有10人陷入濒死境地。同时，在一块满是红树沼泽和丛林的土地上种植庄稼也不是件容易的事。移民者后来回忆说，供给的物资上长满了蛆，有个人说他由于过度劳动，肩膀上全是水疱。他们尝试与当地的库纳人交易，以换取物资，但没成功。不过想想那里恶劣的气候和他们给当地人提供交换的东西——梳子、镜子、假发和羊毛衣服，那些人不感兴趣也就不足为奇了。没多久，苏格兰人就只得靠猎捕大海龟来生存了。

还有一个问题。一道山脉挡住了所有试图通往太平

注释

① contingent [kən'tɪndʒənt] n. 分遣队

② pernicious [pə'nɪʃəs] adj. 很有害的，险恶的

③ unwholesome [ˌʌn'həʊlsəm] adj. 不卫生的，不健康的

④ contagious [kən'teɪdʒəs] adj. 有传染性的，传染病的

⑤ dot [dɒt] v. 散布于

⑥ pre-emptive [pri'emptɪv] adj. 先发制人的，抢先的

⑦ besiege [bɪ'si:dʒ] vt. 包围

⑧ shipwreck ['ʃɪprek] n. 船只失事，（海滩上的）失事船残骸

⑨ calamity [kə'læməti] n. 灾祸，灾难

⑩ ailing ['eɪlɪŋ] adj. 生病的，不舒服的

⑪ bail [beɪl] vt. 帮助脱离困境

⑫ debacle [deɪ'bɑ:kl] n. 崩溃，瓦解

handful of original settlers who had decided to stay on. The new contingent① began trying to rebuild things, without much luck. One of the eventual survivors, a Reverend Francis Borland, described conditions as 'pernicious②, unwholesome③ and contagious④'.

By now the Spanish, whose own new territories were dotted⑤ around Darien, were posing another threat. The English would be no help in this regard. As many embittered Scots would later relate, King William III (King William II of Scotland) wanted to drive the Spanish into the arms of the rival French and would not allow Darien any armed help from nearby Jamaica. English ports and ships were also forbidden to trade with the new Scottish colony. In 1700 some of the settlers at Darien made a pre-emptive⑥ strike against local Spanish forces, with some success, but soon the Spanish were back, besieging⑦ Fort St Andrew. Wracked by disease, the inhabitants were by now dying at the rate of sixteen a day. They held out for a month before surrendering and, in March 1700, those still alive set sail for home. Tragically, on the way back to Scotland, many more died from illness and shipwrecks⑧, with just a handful eventually returning.

Some have suggested that if help had been forthcoming from the English, the Darien colony could have survived and even gone on to be successful. Whatever the reasons for failure, the company had lost a ruinous £232,884 and some 2,000 people had died. Politically, the calamity⑨ had left Scotland weaker rather than stronger. Paterson wrote a tell-all account of the Darien venture in a bid to justify the affair, but he himself became an exponent of the Act of Union; uniting Scotland and England's parliaments. As part of the agreement, the ailing⑩ Scottish economy would be bailed⑪ out. Paterson was eventually granted a pension of £18,000 but the whole debacle⑫

洋的贸易路线。即使有足够多、足够强壮的移民者本可以开辟出这样一条路线。8个月以后，人们决定放弃达连。出发时的1200名移民，只有300人活着回到苏格兰。1699年8月，第二支舰队载着另一批1302名移民驶离苏格兰，当时他们还不知道第一批人已经失败。当他们11月抵达达连时，发现那里几乎已经完全被遗弃，第一批移民中只有少数人坚持留了下来。新来的人开始了新一轮建设，但仍然运气不佳。在最终幸存下来的人当中，有一位叫弗朗西斯·伯兰德（Francis Borland）的牧师，他把那里的环境描述为"邪恶、不健康和传染疾病的"。

此时，西班牙人占领的新地盘已遍布达连周围，这是苏格兰人面临的又一威胁。而英格兰人在这方面没有给予他们任何帮助。后来据许多愤怒的苏格兰人讲述，国王威廉三世（即苏格兰国王威廉二世）想鼓动西班牙与之共同对付法国，并禁止从牙买加附近给予达连任何军事援助。英格兰港口和船只也被禁止与这个新的苏格兰殖民地进行任何贸易。1700年，达连的一些苏格兰人先发制人，对当地西班牙军队发起进攻，取得了一些胜利，但很快西班牙人又进行反攻，包围了圣安德鲁堡。另外，在疾病的摧残下，那里的居民以每天死亡16人的速度减少。他们坚持了1个月，最终投降了。1700年3月，剩下的幸存者起程回家。不幸的是，在回苏格兰途中，又有不少人死于疾病和海难，只剩下少数人最终返回。

有人认为，如果当时英格兰人提供援助，达连殖民地或许能够生存下来，甚至还能成功。但不管什么原因，公司损失了232884英镑，濒于破产，约2000人死于这个

had come at a huge personal price. His own wife and son had perished in Darien.

Today the site of the Darien settlement near the Colombian border is still, by all accounts, a pretty wild place. Its Spanish name, Punta Escoces, or Scottish Point, recalls its ill-fated past. Paterson, however, is remembered on a memorial at the entrance of the Panama Canal. The huge engineering feat, built some 200 years later, provides a shipping route between the Pacific and Atlantic oceans, and is proof that the trading link Paterson imagined was, indeed, achievable.

计划。从政治的角度来说，这个不幸事件使苏格兰变得更加羸弱，而不是强大。帕特森写了一个真实的达连探险经历以解释这件事情，然而，他自己却成为那个将苏格兰和英格兰议会联合起来的《联合法案》的支持者。作为协议的一部分，境况不佳的苏格兰经济将获得帮助。最后帕特森得到了18000英镑的养老金，但是伴随着整个计划的失败，他还承受了个人的巨大损失——他的妻子和儿子在达连殖民地死去。

今天，人们都说，在哥伦比亚边界附近达连移民的居住地仍是一片荒野。它的西班牙语名字叫"Punta Escoces"，意思是苏格兰点，指的就是这段不幸的过去。不过，帕特森的名字被刻在了位于巴拿马运河入口处的纪念碑上。巴拿马运河，这个200年后修建的宏伟工程，在太平洋和大西洋之间形成了一条航运通途，也证实了帕特森所想象的贸易通道确实可行。

~~FAILED~~
THE LOST US STATE OF TRANSYLVANIA

When you mention the word Transylvania it tends to send a shiver down the spine①, conjuring② up images of dark deeds and gory goings on. Of course, most people know that Dracula's legendary home is a real region in today's Romania. But there was once another Transylvania, and its mysterious story is almost as good a yarn③ as Bram Stoker's blood sucking thriller.

The story of the American Transylvania – which could easily have become one of the states we know today – begins in the eighteenth century, just months before the separate colonies broke away from Britain. It was the brainchild④ of a man called Richard Henderson, a judge from North Carolina who aimed to make Transylvania a new colony in its own right, the fourteenth, located where today's Kentucky now sits on the map. Though the European region called Transylvania had existed for centuries, when Henderson set out on his quest⑤ no one had even heard of Dracula. Bram Stoker's novel didn't hit the shelves until 1897.

So why Transylvania? Sylvania is Latin for a forested area. Transylvania basically meant 'on the other sides of the woods'. There was already a colony called Pennsylvania – named after

丢失的美国特兰西瓦尼亚州

每当提到"特兰西瓦尼亚"（Transylvania）这个词的时候，人们常常会觉得脊背凉飕飕的，头脑中会出现一些不光彩的罪行和血淋淋的场景。当然，许多人都知道，**德拉库拉（Dracula）**传说中的家乡位于今天的罗马尼亚。但是曾经有另一个叫特兰西瓦尼亚的地方，关于它的神秘故事几乎与布拉姆·斯托克（Bram Stoker）的吸血恐怖故事一样曲折离奇。

这个美国版的特兰西瓦尼亚故事发生于18世纪，在各个殖民地脱离英国统治数月之前。这个地方本可以轻而易举地成为美国今天的一个州。创建特兰西瓦尼亚州的想法来自一个叫理查德·亨德森（Richard Henderson）的人，他是北卡罗来纳州的一名法官，他想让特兰西瓦尼亚凭借其自身条件成为一个新的殖民地——美国的第14个殖民地，位于今天的肯塔基州。虽然欧洲那个特兰西瓦尼亚已经存在了几个世纪，但当亨德森开始他的尝试时，还没人听说过德拉

注释

德拉库拉是爱尔兰作家布拉姆·斯托克长篇小说《德拉库拉》中的人物。小说描写了吸血鬼——德拉库拉伯爵的施暴和灭亡的故事。其原型是有见血发狂病症的弗拉德三世，出生于特兰西瓦尼亚。

注释

① dubious ['dju:biəs] *adj.* 可疑的
② inauspicious [ˌɪnɔː'spɪʃəs] *adj.* 不吉利的
③ trek [trek] *v.* 长途跋涉
④ outlying ['aʊtlaɪɪŋ] *adj.* 偏僻的，边远的
⑤ constitutional [ˌkɒnstɪ'tjuːʃənl] *adj.* 宪法的，符合宪法的
⑥ delegate [delegate] *n.* 代表，代表团成员
⑦ elm [elm] *n.* 榆树，榆木

its founder William Penn. The territory that Henderson was interested in lay on the other side of the Appalachian mountains, which run north to south down the length of North America's eastern side. Hence Transylvania.

In summer 1774, just two years before the American Declaration of Independence, Henderson formed what would become the Transylvania Company, an audacious and legally dubious① scheme to grab largely wild and unsettled lands to the west and form them into a new state. In March 1775, he managed to buy 81,000 square kilometres from the local Cherokee Indians for an estimated £10,000. He'd also recruited one of America's most famous sons, the explorer Daniel Boone, to help secure the territory. Boone had already attempted to open up new lands in the region and fought local Indians in a short conflict known as Dunmore's War, in which the Shawnee Indians had been forced to give up their claim to the modern region of Kentucky. Now he was asked to blaze a trail into the territory known as the Wilderness Road, along with thirty fellow settlers on what has become a famous journey through the rough terrain. Along the way, three of the party were lost to Indian attacks.

Inauspiciously②, on April Fool's Day 1775, Boone reached a spot on the Kentucky River where he started building a fort and settlement called Boonesborough. Just days later, Henderson trekked③ through the forests to the settlement himself. Though Boonesborough still numbered just 100 people with a few outlying④ settlements, Henderson decided to hold a constitutional⑤ convention to draw up the framework of a government. And, on 23 May, he managed to get delegates⑥ from the wanna-be state to meet under an enormous elm⑦ tree. Amazingly, after three days of debate, an agreement was drawn up which provided for elections, balancing branches

库拉——布拉姆·斯托克的小说直到1897年才上架销售。

那么，为什么要取这个名字？拉丁语西瓦尼亚（Sylvania，也译作"夕法尼亚"——编者注）的意思是"森林地区"。特兰西瓦尼亚的大意是"森林另一头"。那时已经有一个殖民地叫宾夕法尼亚，由它的开拓者威廉·佩恩（William Penn）命名的。亨德森感兴趣的土地位于北美洲东部由北至南的**阿巴拉契亚山脉**的另一侧，所以取名叫特兰西瓦尼亚。

1774年夏，就在美国颁布《独立宣言》两年以前，亨德森成立了一个公司。它是特兰西瓦尼亚公司的前身，打算大面积攫取西部无人居住的荒地，建立一个新的州。这个计划很大胆，而且存在法律问题。1775年3月，亨德森设法以约10000英镑的价钱从当地切罗基族印第安人那里买下81000平方千米的土地。同时，他还请来美国名人、探险家丹尼尔·布恩（Daniel Boone）帮忙获取土地。布恩也打算在这里开辟新的领土。他曾与当地印第安人发生过被称为"邓莫尔之战"的短期冲突，结果肖尼族印第安人被迫放弃了对现在肯塔基州地区的所有权。现在，他要开辟一条进入这个地区的道路——荒野路。他与30个同伴一起，开始了这场著名的穿越这块崎岖之地的历险。一路上，他们中有3个人在印第安人的进攻中丧生。

1775年4月1日愚人节这天（这听上去有点不吉

注释

阿巴拉契亚山脉位于美国东部，是北美洲东部众多山脉的统称，又称为阿巴拉契亚高地或阿巴拉契亚山系。

注释

《独立宣言》是北美洲十三个英属殖民地宣告自大不列颠王国独立，并宣明此举正当性的文告。1776年7月4日，该宣言由第二次大陆会议（Second Continental Congress）于费城批准。

注释

① petition [pə'tɪʃn] *vi.* 祈求，请求
② utopian [ju:'təʊpɪən] *adj.* 乌托邦的，不切实际的
③ crony ['krəʊni] *n.* 密友，亲密的伙伴
④ lambaste [læm'beɪst] *vt.* 严厉斥责某人
⑤ secede [sɪ'si:d] *v.* 退出，脱离（组织等）

of government and even courts; a system not too dissimilar to the US constitution, which, incidentally, was not adopted until 1787.

Without wasting any time, Henderson and the rest of the Transylvania Company's owners then petitioned① the Continental Congress to become the fourteenth colony, stating that they were 'engaged in the same great cause of liberty'. John Adams, who in 1776 would be a founding father of the independent United States, warmed to the Transylvania investors, saying that they were 'charged with republican notions – and utopian② schemes'. But George Washington, the first US president said: 'There is something in the affair which I neither understand, nor like, and wish I may not have cause to dislike it worse as the Mistery [*sic*] unfolds.'

There were other big problems. The colonies of both Virginia and North Carolina already had claims on the territories, and many of the settlers decided they weren't keen on Henderson's methods. He and his Transylvanian cronies③ were lambasted④ as blood suckers who merely wanted the land for their own economic gain. Henderson was soon forced to back down. In 1778 the Virginian Assembly declared the Transylvania claim void for good. Henderson himself didn't come off too badly, being awarded 200,000 acres as compensation for his efforts to colonise the West. His proposed capital of Boonesborough, however, was frequently attacked by Indians over the coming years while Boone himself was left almost penniless thanks to a robbery and some failed business ventures.

Henderson's ambitious attempts at state building are no means unique in American history. There are plenty of other bizarre states that never made it on to the familiar list of fifty that we know today. In 1850 a mining town in California called Rough and Ready tried to avoid taxes by seceding⑤ from the

利），布恩到达肯塔基河，开始在那里修筑堡垒，安营扎寨，并把那里命名为"布恩斯巴罗"。仅几天后，亨德森本人穿过森林跋涉到这里。尽管布恩斯巴罗和几个边远的安置点总共只有100人，但亨德森仍决定召开制宪会议，拟订政体框架。5月23日，他组织那些想成立这个州的代表聚在一棵大榆树下开会。令人惊奇的是，只经过3天讨论，他们就制定了一项协议，包括选举程序、相互制衡的政府机构，甚至还有法庭制度，这个体制与美国宪法并无太大差异。顺便说一句，美国宪法是1787年制定的。

接着，亨德森和特兰西瓦尼亚公司的其他股东向大陆国会申请成为第14个殖民地，声称他们"投身于同样伟大的自由事业"。后来成为1776年美国独立的开国元勋之一的约翰·亚当斯（John Adams）对特兰西瓦尼亚的投资者产生了好感，说他们"充满共和的思想，还有乌托邦的计划"。但是美国第一任总统乔治·华盛顿却说："对于这件事，有些东西我既不理解，也不喜欢，希望当这个谜团（原文如此）解开的时候，我不会更不喜欢它。"

还有一些较大的问题。那时，弗吉尼亚州和北卡罗来纳州已经声称拥有这个地区，而许多殖民者对亨德森的路子也不感兴趣。他和他的特兰西瓦尼亚那伙人被斥为"只想通过这块土地为自己攫取经济利益的吸血鬼"。不久，

▲ 关于1775年5月23日在布恩斯巴罗召开的特兰西瓦尼亚州制宪会议的一幅无署名草图。这个未能成立的州现在是美国肯塔基州的一部分。

An anonymous sketch showing the Transylvania Constitutional Convention at Boonesborough, 23 May 1775. The abortive US state is now part of modern day Kentucky.

注释

① budding ['bʌdɪŋ] *adj.* 开始发展的，成长中的
② vampire ['væmpaɪə(r)] *n.* 吸血鬼

US. In the early part of the twentieth century residents of Oklahoma and Texas, mad at the lack of roads on which to drive Henry Ford's new cars, tried to set up Texlahoma. A similar bid in the 1970s was made by Forgottonia, a region in western Illinois which also believed it was missing out on services and development. Eventually all these attempts went the same way as Transylvania.

All is not lost for Dracula fans though. The idea of the American Transylvania lived on. Kentucky's Transylvania University, where budding① vampires② would surely want to study, was founded in 1780, named after the failed state. And, in 1861, a county in the state of North Carolina got the name – as a tribute to the original company.

亨德森就被迫作出让步。1778年，弗吉尼亚州议会宣布特兰西瓦尼亚州永久无效。亨德森本人的结局还不算太糟，他获得了约8万公顷（20万英亩）土地的奖励，作为他在西部开拓殖民地努力的补偿。他提议的首府布恩斯巴罗却在之后的岁月里频频遭到印第安人的攻打，布恩自己也因为一次被劫和一些失败的投资而变得身无分文。

在美国历史上，不仅只有亨德森雄心勃勃地尝试过创立一个州，还有许多其他光怪陆离的州最终没能被列入今天我们所知的50个州。1850年，加利福尼亚州拉夫安雷迪市的一个采矿镇想通过脱离美国以逃避纳税。20世纪初，俄克拉何马州和得克萨斯州的居民，由于没有公路供他们驾驶亨利·福特（Henry Ford）的新车而非常恼火，想成立得克拉何马州。20世纪70年代，伊利诺伊州西部一个叫**弗格托尼亚**的地区也发生过类似的事情，那里的人们认为没有得到应有的服务，错过了发展机会。最终，所有这些尝试都像特兰西瓦尼亚一样失败了。

不过，对于喜欢德拉库拉的人来说，他们还没有失去一切，美国特兰西瓦尼亚思想存留了下来。位于肯塔基州的特兰西瓦尼亚大学成立于1780年，以这个没能成立的州的名字命名。那里一定是小吸血鬼们想去学习的地方。1861年，北卡罗来纳州的一个县也取了这个名字，算是对最初那个公司的纪念吧！

校订者注
弗格托尼亚源自英文"被遗忘的"一词。

ABANDONED
THE FRENCH REPUBLICAN CALENDAR

注释

① dispense with sth. 摒弃，摆脱某事
② metric ['metrɪk] *adj.* 十进制的
③ decimalise ['desɪməlaɪz] *v.* 将（货币）改为十进制，以小数表示一数字

When the French people took control of their destiny and dispensed① with the services, and indeed the heads, of their ruling classes following the 1789 revolution, much was done to erase the signs and symbols of the past. In came a new system of government. In came a new way of measuring distances and weights. And, to the puzzlement of the masses, in came new days of the week, months of the year, and, most baffling of all, the way of measuring time itself. For in the new French Republic, history was consigned to the history books. 1789 became year one: the start of a brand new calendar based on a much simplified, logical metric② system. This was decimalised③ time.

One hundred seconds made a new minute, 100 minutes made an hour, and ten hours made a day. Each new French week consisted of ten days, the ten-day cycle becoming known as a *décade*, and three *décades* made a month. The rule of ten then descended into slight chaos as twelve thirty-day months formed the year, but at least this unified months into regular periods and ended the nonsense of some months consisting of thirty-one days, others thirty, and one with twenty-eight or twenty-nine.

法国共和历

1789年大革命之后，法国人民掌握了自己的命运，除掉了统治阶级的压迫——包括他们的头，做了大量工作清除过去的符号和标志。他们创立了新政体，创立了新的度量衡制度，而令人困惑的是他们还创立了新的日、周、月、年乃至时间的测量方式。因为对于新的法兰西共和国而言，历史应该被丢弃在历史书里。1789年成为元年：成为以一种相当简单、具有逻辑的度量方式为基础的全新历法的开始。这就是十进制时间。

100秒组成新的1分钟，100分钟组成1小时，10小时组成1天。每个新的法国周由10天组成，10天1周又称为旬，而3旬组成1个月。这种十进制在由12个30天的月组成一年时稍微出现了一点混乱，但这样至少统一了各月的天数，结束了有的月31天、有的月30天，还有一个月一会儿28天一会儿29天的那些无聊规定。尽管法国共和历的主要设计师数学家吉尔贝·罗默

注释

① precedent ['presɪdənt] *n.* 前例，先例
② dogma ['dɒgmə] *n.* 教义，教条
③ regime [reɪ'ʒiːm] *n.* 政治制度，政权
④ alter ['ɔːltə(r)] *v.* 改变，更改
⑤ axis ['æksɪs] *n.* 轴，轴线
⑥ inextricable [ˌɪnɪk'strɪkəbl] *adj.* 无法分开的，不能逃避的
⑦ Catholic saints' days 天主教的圣日

Despite the logic with which mathematician Gilbert Romme, the main architect of the French Republican Calendar, had designed the system – having based it on the experience on old Egyptian and Athenian calendars – precedent① wasn't on his side. Since time began, calendars had traditionally been linked with astronomy. But the earth, moon and sun were no respecters of digital dogma②, leading to the new calendar being five and a bit days light of the standard solar year – the 365 days, 5 hours, 48 minutes and 56 seconds it takes for the earth to circle the sun.

If the new republican regime③ could have altered④ the spinning of the earth on its axis⑤, no doubt they would have tried. Instead, Romme and his republican committee colleagues who had been set the task of redesigning time, had to compromise. With regret, five extra days were added to the end of every year. Worse, leap years were to remain; this time tagged onto the end of the year, rather than in February (which itself now spanned two new months called wind and rain). New leap years were scheduled every four years from the end of year three (so three, seven, eleven and so on), a further deliberate attempt to break with pre-Revolutionary tradition.

In line with the new age of reason, culture, ideology and science, poet Philippe François Nazaire Fabre, known as Fabred'Eglantine, was chosen to select names for the days and months. He chose to honour fruit, vegetables, animals and even pieces of agricultural equipment for every one of the 366 days. The old calendar's system of holidays, inextricably⑥ linked with Catholic saints' days⑦, even Christmas, were replaced with a series of days at the end of the year thanks to the five lunar days that Romme couldn't avoid. These celebrated different liberties, including one that allowed Frenchmen to libel one another without fear of legal retribution – quite something when someone could be condemned to the guillotine for anti-

（Gilbert Romme）设计这一历法所用的逻辑是以古埃及历法和雅典历法的经验为依据的，但引用这些先例并非对他有利。自纪元以来，历法在传统上都与天文联系在一起。但地球、月亮和太阳并不遵守数学规律，导致新历法比标准太阳年（地球绕太阳一圈的时间为365天5小时48分56秒）少了5天多。

假如新的共和制能够改变地球沿轴线的转动，那么他们一定会试一试。然而，罗默和重新设计历法的共和派同志不得不采取折中手段，满怀遗憾地在每年的最后加了5天。更糟糕的是仍然有闰年，这次把多余的那天放在了年终，而没有放在2月（2月本身则被分成了风月和雨月）。为了刻意打破革命之前的传统，他们从第三年年末起每四年安排一次新闰年（于是3年、7年、11年都是闰年，以此类推）。

为了与新时代的理性、文化、意识形态和科学技术保持一致，选定诗人菲利普·弗朗索瓦·纳泽尔·法布尔（Philippe François Nazaire Fabre，人称"蔷薇法布尔"）为新历法的日、月择名。他选择了用水果、蔬菜、动物甚至农业设备的名字来为366天的每一天命名。旧历法的节日都与天主教的圣日甚至圣诞节密切相关，现在都被年末的那几天（罗默无法避免的日子）取代了。这些日子用于庆祝不同的解放运动，包括一个允许法国人相互控告而不必担心受到法律惩罚的节日——这实在是个美妙的日子，若非在这样的日子，同情反革命的人就要被送上断头台了。

revolutionary sympathies.

Traditional Calendar	Republican Calendar	Meaning
September / October	vendémiaire	Wine harvest
October / November	brumaire	Mist
November / December	frimaire	Frost
December / January	nivôse	Snow
January / February	pluviôse	Rain
February / March	ventôse	Wind
March / April	germinal	Germination
April / May	floréal	Flowering
May / June	prairial	Meadow
June / July	messidor	Harvest
July / August	thermidor	Heat
August / September	fructidor	Fruit

The five new holidays at the end of the year, beginning the day after 30 fructidor:

1 Fete de la vertu (holiday of virtue)

2 Fete du genie (holiday of genius)

3 Fete du travail (holiday of work)

4 Fete de l'opinion (holiday of opinion and free speech)

5 Fete des recompenses (holiday of rewards)

6 Jour de la revolution (day of the revolution) (the leap day)

With numbers and names decided, all that was now needed was to select a start date, and here too anxious debate raged[①]. Some revolutionaries believed that it should coincide[②] with the overthrow of the monarchy – the Revolutionary Calendar – others that it would be more fitting for it to commemorate[③] the new republic, which happily coincided with the autumn equinox[④], on 22 September 1792. The second option won the day and the calendar was formally adopted on 24 October 1793 (backdated to 22 September the previous year), although because getting

注释

① rage [reɪdʒ] v. 发怒，怒斥
② coincide [ˌkəʊɪn'saɪd] v. 与……一致，想法、意见等相同
③ commemorate [kə'meməreɪt] v. 纪念，庆祝
④ equinox ['iːkwɪnɒks] n. 昼夜平分点，春分或秋分

传统历法与共和历的对比

传统历法	共和历	意思
9月/10月	葡月	收获葡萄酒
10月/11月	雾月	起雾
11月/12月	霜月	降霜
12月/1月	雪月	降雪
1月/2月	雨月	下雨
2月/3月	风月	刮风
3月/4月	芽月	发芽
4月/5月	花月	开花
5月/6月	牧月	放牧
6月/7月	获月	收获
7月/8月	热月	热
8月/9月	果月	水果

年末的五个新节日，在果月30日之后，分别是：道德日、才能日、工作日、舆论日、奖赏日。遇闰年再加上革命日（闰日）。

确定了日期和名字之后，还需要选择一个起始日，这下引起了激烈争论。有些革命者认为，革命历的起始日应该与废除君主政体的日子保持一致；另一些革命者认为，为了纪念新的共和国，起始日更适合与秋分（1792年9月22日）保持一致，因为这一天恰是共和国成立日。最终第二种方案占了上风，于是1793年10月24日正式采用了新历法（追溯至前一年的9月22日）。不过由于18世纪时把这一消息传遍法国各地尚需时日，所以法国各地采用共和历的时间有所不同。

地球沿椭圆轨道绕太阳运行，这就使得事态更加复杂了。根据直到当时乃至现在使用的格里高利历，秋分一般为9月22日或9月23日。显然，对于一项逻辑严密的历法制

注释

① elliptical [ɪ'lɪptɪkl] *adj.* 椭圆的
② lunacy ['lu:nəsi] *n.* 精神失常，精神错乱
③ ruthlessly ['ru:θləslɪ] *adv.* 无情地，冷酷地，残忍地
④ henceforth [ˌhens'fɔ:θ] *adv.* 从今以后，今后
⑤ synchronise ['sɪŋkrənaɪz] *v.* （使）同步，（使）同速进行
⑥ reckoner ['rekənə] *n.* 计算器
⑦ decimal ['desɪml] *adj.* 十进位的，小数的

the message around the country from Paris took a little time in the eighteenth century, different parts of the country began to use the calendar at different times.

The planet's elliptical① journey round the sun caused a further complication. Under the Gregorian calendar that had been used until then and which we use today, the equinox can fall on either 22 or 23 September. Clearly, for a logical system, this is utter lunacy② and on this matter Romme was not going to give in to physics. He reasoned that once the Republican Calendar was introduced, provided it was ruthlessly③ enforced, 22 and 23 September would be an irrelevance. Henceforth④ the new date of I harvest would synchronise⑤ the start of the New Year and the equinox.

The period from what is now 22 September 1792 to 21 September 1793 became year one (expressed as roman numeral I), presenting a further complication; one not experienced for 1792 years. History had to be re-dated. Calculations were made to show that the Creation occurred 5,800 in advance of the Republic, Noah built his ark 4,144 years previously, and printing had been invented 330 years before. In year VIII of the new era, an almanac of practical information for the household – a common publication in French districts of the period – looked to the future too, providing a ready reckoner⑥ that helped people calculate the average number of years they had left to live at any particular age.

With the calendar up and running, the authorities waited for the people to take it to their hearts. But it only caused trouble, confusion and frustration. Decimal⑦ weeks now meant that weekends fell every ten days, instead of every seven. Weekly markets, events, meetings, even paydays suddenly became less frequent. And hours almost doubled in length, which made working days disagreeably long for the French. With all these

度来说，这简直太荒谬了。在这个问题上，罗默不打算再向物理规律低头。他认为只要引入共和历，只要严格执行这一历法，那么就与秋分是在9月22日还是在9月23日没有多大关系。因此，获月的第一日就与新年元旦和秋分同步开始。

这样，1792年9月22日至1793年9月21日就成了元年（用罗马数字Ⅰ表示），结果事态变得空前复杂。这是之前1792年里都没遇到的问题——史书上的日期必须进行改写。经过一番计算，创世发生于共和前5800年，诺亚方舟建于共和前4144年，而印刷术发明于共和前330年。在共和历Ⅷ年，又出版了一本面向未来的家用年历，这是一种当时在法国常见的出版物，帮助各个年龄段的人计算他们还能活的平均年数。

随着共和历的逐步推广，政府等待人民打心眼儿里接受它，然而它带来的只有麻烦、混乱和沮丧。十天一周意味着十天才有一次周末，而不是七天。周市、周赛、周会，甚至发薪月也突然变得频率低了，而小时的长度几乎长了一倍，这让法国人的工作日变得十分漫长，甚至令人难以接受。有了这些新名字之后，许多人根本不知道一周中的哪天是什么意思，而占多数人口的天主教教徒发现取消礼拜天特别让人觉得匪夷所思。**普瓦图-夏朗德大区**普勒马丹村的青年男女按传统一般在周三成婚，他们发现新指定的周三每十天才出现一次，这让婚姻登记处倍感压力。

不满孕育着反叛，但试图逃避共和历的努力都遭到了无情的镇压。警察突袭关闭了一些在非法定日期或时间营

注释

普瓦图-夏朗德大区是法国西部的规划大区，共有9个省，7座人口超过10万的城市，总人口500万，相当于明尼苏达州或法国普罗旺斯-阿尔卑斯-蓝色海岸地区的人口总数。

注释

① fathom ['fæðəm] v. 充分理解、领悟

② Poitou-Charentes [pwaː'tuː ʃɑː'rɔŋt] n. 普瓦图-夏朗德（法国西部的规划大区）

③ circumvent [ˌsɜːkəm'vent] v. 设法克服或避免

④ swoop [swuːp] v.（鹰）俯冲，猛冲

⑤ guillotine ['gɪlətiːn] n. 断头台

⑥ align [ə'laɪn] vt. 使成一线，使结盟，排整齐

new names, many people quite literally didn't know what day of the week it was, and the still largely Catholic population found the abolition of Sundays particularly hard to fathom①. Pleumartin, a village in Poitou-Charentes② where couples traditionally married on Wednesdays, discovered that the newly assigned day Quintiti turned up only every tenth day, putting a pressure on slots at the register office.

Discontent bred rebellion, but attempts to circumvent③ the calendar were ruthlessly repressed. Police swooped④ to stop markets being held on the wrong days or at the wrong times, and even to enforce the new day of rest, the décadi. Brave traders and shoppers sometimes fought back, attacking gendarmes, but this was risky. At the height of the French Terror, anything that could be interpreted as anti-revolutionary could be severely punished. With wonderful irony, Romme himself eventually fell victim to the Terror. At the time of the revolution he had voted for the king's execution, but since then he had made protests against the new regime. Arrested, tried and sentenced to death, he committed suicide on his way to the guillotine⑤ in 1796, using a small knife he had secreted about him rather than wait for the big chop. At this, he became the Martyr of Prairial – prairial being the ninth month of the Republican Calender.

The calendar didn't survive much longer than its chief architect. Sensing its unpopularity, Napoleon later agreed an accord with the Pope, and at midnight on 31 December 1805 (10 Nivôse, year XIV), time was called on the French Republican Calendar. After nearly thirteen years, the country aligned⑥ its clocks with the rest of the world on 1 January 1806. This wasn't quite the end. The calendar enjoyed a very brief revival later in the century when the Paris Commune government, which briefly ruled the capital in 1871, restored it for ten weeks between 18 March and 28 May.

业的市场，甚至强制执行新的休息日——周十。勇敢的商人有时会予以回击，甚至袭击宪兵队，但这样风险太大。在法国恐怖的高峰期，凡是可以被定性为"反革命"的行为都会遭到严惩。具有讽刺意味的是，罗默本人也成了恐怖的受害者。在革命时，他投票赞成处决国王，但从那时起他又开始抵制新政权。于是他遭到逮捕和审判，最终被判处死刑。他于1796年在通往断头台的路上用一把私藏的小刀自杀，否则等待他的便是铡刀了。这样，他成了牧月（共和历的9月）的殉道者。

作为共和历主要设计者的罗默死后，共和历本身也好景不长。不久之后，拿破仑意识到共和历不受欢迎，就与教皇达成协议，在1805年12月31日午夜（共和历XIV年雪月连枷日）叫停了法国共和历。时隔大约13年之后，法国于1806年1月1日将其时钟再次与世界其他国家的钟表"对时"。不过到此并未结束，19世纪末巴黎公社政府于1871年统治巴黎时，共和历再次回光返照，在3月18日至5月28日之间又恢复了10周。

FAILED
LATIN MONETARY UNION

注释

① indicator ['ɪndɪkeɪtə(r)] *n.* 指示器，[化]指示剂
② banish ['bænɪʃ] *vt.* 放逐，排除（想法等）
③ consolidation [kənˌsɒlɪ'deɪʃən] *n.* 巩固，联合，合并
④ initiative [ɪ'nɪʃətɪv] *n.* 主动性，积极性

If nineteenth-century history is anything to go by, changing a workable system that is universally understood and with which most people appear happy was becoming a French characteristic. For what they had already done successfully for measures, but disastrously for dates, French republicans now attempted with currency. In short, out with the old, in with the new. Gold – the international indicator① of national wealth and the backbone of banking – would no longer be permitted in French coins from 1803.

To 'banish② gold', as one historian called it, was thought by many economists to be madness, especially at the start when the Napoleonic wars needed to be paid for. But once the wars ended and the French economy returned swiftly to its feet, neighbouring countries started to look on with increasing admiration. Starting to base their own economies on the franc, a currency consolidation③ began that would last throughout the middle 1800s. Soon a formal monetary union was underway, and France was at its heart.

This initiative④, called the Latin Monetary Union (LMU), had its origins in Europe's changing political geography and

拉丁货币同盟

　　如果19世纪的历史可以作为依据的话，那么改变已被大众普遍理解并让大部分人都满意的有效可行的制度已成为法国的一个特征。他们成功修改了度量衡，但修改历法失败了，于是法国共和派又准备拿货币开刀——总之就是革故鼎新。黄金作为国家财富和银行实力的国际象征，自1803年起在法国禁止作为货币使用。

　　许多经济学家认为废除黄金的做法近乎疯狂，特别是在拿破仑战争开始时需要黄金付款的情况下。但战争一结束，法国经济迅速回升，邻国开始对其越来越羡慕。邻近国家开始把经济建立在法郎基础之上，开启了一轮持续到19世纪中叶的货币整合。不久之后，欧洲成立了一个正式的货币同盟，法国是其核心。

　　这个同盟名为"拉丁货币同盟"，起源于欧洲不断变化的地缘政治和不断增长的经济稳定需求。比利时于1830年脱离荷兰，在之后的两年内，比利时逐渐放弃荷兰盾，转而使用法郎。意大利在统一之路上也在经历着一系列似

注释

① ditch [dɪtʃ] vt. 摆脱，抛弃

② stagger ['stægə(r)] v. 蹒跚，犹豫，动摇

③ manipulate [mə'nɪpjuleɪt] vt. 操纵

④ minted ['mɪntɪd] adj. 崭新的，新制作的

⑤ parity ['pærəti] n. 平价，价值对等，同等

⑥ bimetallism [baɪ'metlɪzəm] n. 复本位制（尤指金银二本位制，在货币的法定纯度中保持二者的固定比率）

⑦ inception [ɪn'sepʃn] n. 开始，开端，初期

⑧ denomination [dɪˌnɒmɪ'neɪʃn] n. 宗派，教派

⑨ ratio ['reɪʃiəʊ] n. 比，比率

the quest for economic stability across the continent. Within two years of Belgium separating from Holland in 1830, it had ditched① the Dutch guilder for francs. Italy, going through a seemingly never-ending series of revolutions on its way to unification, aligned with the franc in 1861. With other countries happy to link to a single currency too, a path was set for a more formal union, with Spain, Greece, Romania, Bulgaria, Venezuela, Serbia and the Vatican going Francophile a year after its introduction in 1866. The currency union staggered② on for fifty years, but was manipulated③ by a misjudgement about the value of gold relative to silver years before it even began.

Back in 1803 France decided that all newly minted④ coins, up until then a mix of gold and silver, should be made out of silver only. At this time, coins in many countries, including France, contained real gold and silver in carefully regulated amounts, based both on the weight of the precious metal and its fineness. And because under a system called 'free coinage' anyone could have their own gold and silver pressed into coins, it was crucial that the value between the two metals was correctly established. If there wasn't 'parity⑤' between the two, one could be used to buy the other very cheaply which could be then sold elsewhere for a profit. This interchange between silver and gold was known as bimetallism⑥, and because the parity between the two metals was wrong at the start, the Latin Monetary Union was arguably doomed from inception⑦.

LMU experts Kee-Hong Bae and Warren Bailey explain:

The weight, fineness, and denomination⑧ of coins defined a 'mint ratio⑨' for the value of gold versus silver. If, for example, a silver one-franc coin contained five grams of silver and a gold twenty-franc coin contained six and two-thirds grams of gold, the implied mint parity was 15 to 1.

乎永无止境的革命，也于1861年开始使用法郎。其他国家也愿意采用单一货币，这为成立一个更正式的同盟铺平了道路。西班牙、希腊、罗马尼亚、保加利亚、委内瑞拉、塞尔维亚和梵蒂冈陆续在该同盟1866年开始实行1年后加入货币同盟，转向了亲法阵营。货币同盟蹒跚而行50年，但在它成立之前就始终受黄金相对白银价值的错误估计所掌控。

追溯至1803年，法国决定所有新铸硬币只得用银铸成，而在此之前一直是用金银混合铸成的。这时，包括法国在内的许多国家的硬币都含有精确定量的真金、真银，分量主要取决于贵金属的重量与成色。而在货币自由铸造制度下，任何人都可以把自己的金银压成硬币，所以准确确定这两种金属之间的价值非常重要。如果金银之间不存在平价，则人们可以用一种金属非常便宜地购买另一种金属，然后再卖到其他地方盈利。金银之间的这种交易被称为复本位制，而由于这两种金属之间的平价在开始时不正确，所以拉丁货币同盟可以说从一成立就注定是失败的。

拉丁货币同盟研究专家斐基洪（Kee-Hong Bae）和沃伦·贝利（Warren Bailey）解释道：

> 硬币的重量、成色和面额决定了金对银价值的铸造比率。例如，假设一枚1法郎银币含5克银，一枚20法郎金币含 $6\frac{2}{3}$ 克金，则铸造比率为15:1。

注释

① germinal ['dʒɜ:mɪnəl] *adj.* 幼芽的

② bimetallic [ˌbaɪmɪ'tælɪk] *adj.* 二金属的，复本位制的

③ transaction [træn'zækʃn] *n.* 交易，业务，事务

④ alloy ['ælɔɪ] *v.* 合铸，熔合（金属）

⑤ formalise ['fɔ:məlaɪz] *v.* 使成为正式的

⑥ rebuff [rɪ'bʌf] *vt.* 断然拒绝

If this was complex, it didn't matter. All people really cared about was whether their money kept its value. Nevertheless, 15:1 was a crucial ratio, but it wasn't the one that the French set when they changed their currency in 1803. On 7 germinal① XI in the French Republican Calendar, the country started producing new francs containing 5 grams of silver 0.900 fine; a bimetallic② ratio that slightly overvalued silver at 15.5:1. This was a mistake that was to prove costly further down the line. Furthermore, although France stopped producing most gold coins, old ones remained legal tender and a new coin for high value transactions③, named the 'germinal franc' after the month of germinal in the Republican Calendar, was pressed to mark the occasion.

As time went by and other countries joined the francozone, each nation decided on the fineness of silver in its own coins. In Italy, 0.835 of silver to alloy④ was agreed, while Switzerland chose 0.800. As monetary union took a step closer, every country wanted its own fineness to become the adopted standard to save it having to mint new coins. France, most of all, was determined to get its way and reduced the fineness of the silver in its coins to 0.835 in 1864, just before the introduction of the union. Other countries followed suit. Eight coins were issued throughout the four founder nations of France, Belgium, Italy and Switzerland, when the Latin Monetary Union was formalised⑤ in August 1866.

The United Kingdom, true to form, resisted, saying while it would like not to be excluded, the LMU was a continental idea and the UK would prefer instead to go it alone and concentrate on strengthening the pound. Changing its mind within two years, the chancellor of the exchequer, Robert Lowe, tried to agree a deal with France to join, but was rebuffed⑥. Even so, some key figures in the UK were being turned onto the idea as

如果这个解释比较复杂，那没关系。所有人真正关心的是他们的钱是否保值。无论如何，15:1是个关键比率，但法国人在1803年改变币制时却没有采用这个比率。法国共和历XI年芽月桦树日，法国开始生产含5克成色为0.900银的新法郎，铸造比率略偏重于银，达15.5:1。这一错误接下来被证明代价惨重。此外，虽然法国已经停止了生产大部分金币，但旧金币仍然是法定货币，另外为了纪念这一事件，法国又铸造了一种以共和历芽月命名的新币——芽月法郎，用于大额交易。

随着时间流逝，其他国家也加入了法郎区，各国分别规定了各自硬币中银的成色。意大利规定银的成色为0.835，而瑞士选择了0.800。随着货币同盟进一步发展，各国都想使其规定成色成为普遍采用的标准，以免铸造新币。更糟糕的是法国决定自行其是，于1864年将银的成色减至0.835，这时货币同盟马上就要成立了。其他国家紧随其后。当拉丁货币同盟于1866年8月正式成立时，法国、比利时、意大利、瑞士这4个创始国共发行了8种硬币。

英国一如既往地表示反对，称其虽然不愿意被排除在外，但拉丁货币同盟是整个欧洲大陆的事，英国宁愿单干，把精力集中在强化英镑上。不到两年英国又变卦了，财政大臣罗伯特·洛韦（Robert Lowe）尝试与法国谈判加入货币同盟，但遭到断然拒绝。纵然如此，英国的一些关键人物还是始终把这一想法作为当时议会正在争论的十进制化的逻辑延伸。资深银行家和执政自由党政府的顾问奥弗斯通（Overstone）勋爵提议1英镑等价于25法郎。他认为还应引入一种称为"女王"的20法郎新硬币，代替1英

a logical extension to decimalisation① that was being debated in parliament at the time. A senior banker and an advisor to the ruling Liberal government, Lord Overston, proposed that the pound be equalised to 25 francs. A new 20 franc coin should be introduced too, to be known as a 'Queen' to replace the sovereign② he said. Overston's idea was laughed out of parliament, though he persisted in his case at home and abroad.

Although the UK was on the sidelines③, eighteen countries eventually joined the LMU. But the rules were complicated and inconsistent, and there were too many loopholes④. For economic reasons, a cap was placed on the amount of money each nation could produce: no more than six franc coins for each person in their country. And each member could strike low-value silver coins containing less silver than their face value for use in their country alone. Under LMU regulations, government bodies and local authorities were obliged⑤ to accept silver coins from any country in the monetary union up to a value of 100 francs per transaction. With differing amounts of silver in the coins, people flocked to change low-ratio silver coins to ones with higher silver content, causing silver to become overvalued. Whole countries took advantage. Coins from nations with low silver content were exchanged in other countries for gold. Germany, not part of the union, found a rich seam⑥ in sending officials to Paris with silver ingots where they would be minted into five franc coins, which were exchanged first for banknotes⑦, then for gold. Even the Pope was at it. The papal⑧ state's treasurer, Giacomo Cardinal Antonelli, ordered coins to be minted that were light on silver and then exchanged them for coins from other countries that had the correct amount of the metal.

France soon experienced a run on its iconic Napoleon 22-carat gold coins. At the time, a gold Napoleon had a face value of 20 francs. People found they could mint four 5-franc

注释

① decimalisation [ˌdesɪməlaɪ'zeɪʃən] *n.* 十进制
② sovereign ['sɒvrɪn] *n.* 君主，最高统治者
③ sideline ['saɪdlaɪn] *n.* 副业，兼职
④ loophole ['lu:phəʊl] *n.* 漏洞
⑤ oblige [ə'blaɪdʒ] *v.* 强迫或要求某人做某事
⑥ seam [si:m] *n.* 接缝，接合处
⑦ banknote ['bæŋknəʊt] *n.* 钞票，纸币
⑧ papal ['peɪpl] *adj.* 教皇的

镑的金币。奥弗斯通勋爵遭到了议会的冷嘲热讽，但是他在国内外仍然一贯坚持自己的提案。

虽然英国一直在袖手旁观，但先后有18个国家加入了拉丁货币同盟。不过由于规则太复杂而且前后不一致，所以存在许多漏洞。出于经济原因，同盟对各国生产的硬币数量作了限制：人均不得超过6法郎硬币。同时各个成员国可以铸造含银量少于面值的不足值银币，仅限于国内使用。根据拉丁货币同盟规定，政府机构和地方政府有义务在每次交易中接受来自货币同盟内任一国家高达100法郎的银币。由于银币中含银量不同，人们纷纷把低成色银币换成高成色银币，造成银价过高。所有国家都开始占便宜，来自低成色国家的银币在其他国家被换成黄金。德国虽然不属于货币同盟，但也发现了发财机会。他们派一些官员带银锭前往巴黎，在巴黎铸成5法郎硬币，先换成钞票，再换成黄金。即使教皇也乐此不疲，教皇的财务大臣枢机主教贾科莫·安东内里（Giacomo Cardinal Antonelli）命令把硬币中银铸得少些，然后换成其他国家含有足额银的硬币。

法国很快出现了22克拉拿破仑头像金币挤兑潮。当时，1枚拿破仑金币面值为20法郎。人们发现可以用价值16法郎的白银铸4枚5法郎的银币，买上1枚拿破仑金币就可以赚到差价。1853—1857年，法国损失了11.25亿法郎白银。

直到1874年，金银兑换率的操纵仍然经久不衰，当时货币同盟暂停了金银自由兑换，开始坚持金本位制度（这是一种英国于1816年采纳的国际货币制度，以固定重量的

注释

① manipulation [məˌnɪpjʊˈleɪʃn] n. (熟练的) 操作，控制

② unabated [ˌʌnəˈbeɪtɪd] adj. 不减弱的，不减退的

③ convertibility [kənˌvɜːtəˈbɪlɪtɪ] n. 兑换

④ humiliate [hjuːˈmɪlieɪt] v. 使蒙羞，使丢脸，使出丑

⑤ demise [dɪˈmaɪz] n. 死亡，(不动产的) 转让，让位

⑥ cease [siːs] v. 停止，终止，结束

coins with 16 francs worth of silver, buy a Napoleon and make an immediate profit. During the five years between 1853 and 1857, France lost 1,125 million francs of silver.

The manipulation① of the exchange rate between gold and silver continued unabated② until 1874, when the union suspended the convertibility③ of gold to silver and stuck to the gold standard, the international monetary system based on fixed weights of gold that the UK had adopted in 1816. For all the countries in the LMU, it was a humiliating④ climb down, if not quite the end for the LMU. After four years, the temporary suspension became permanent – and although no new silver coins were struck in the LMU, the old ones continued to be legal tender. The widening gap between the value of gold and silver and the impact of the Great War made survival of a currency union almost impossible. Although the LMU's formal demise⑤ did not come for a further fifty years – until 1926 – in practice, silver-light francs exchangeable freely across eighteen nations had ceased⑥ to exist long before.

金为基准）。对于拉丁货币同盟的所有国家来说，这是一次耻辱的退让，4年之后，暂停变成了永久停止。虽然拉丁货币同盟不再铸新银币了，但旧银币仍然是法定货币。金银之间的差价不断扩大，之后受第一次世界大战的影响，货币同盟已名存实亡。虽然拉丁货币同盟的正式消亡是在其成立50年之后的1926年，但18国之间不足值银币的自由兑换实际上早就停止了。

A TAX ON LIGHT AND AIR

注释

① entrepreneur [ˌɒntrəprə'nɜ:(r)] *n.* 企业家，主办人
② buck [bʌk] *n.* （美）钱，元
③ clipped coins 削边货币，削边指外郭，即钱身外周之突出部分向外斜削，与额轮相反
④ inflation [ɪn'fleɪʃn] *n.* 膨胀，通货膨胀
⑤ dire ['daɪə(r)] *adj.* 可怕的，恐怖的
⑥ two-pronged 双交叉的
pronged [prɒŋd] *adj.* 尖端分叉的
⑦ incur [ɪn'kɜ:(r)] *v.* 招致，引起
⑧ stamp [stæmp] *v.* 盖章于……，标记

Coins that were light on silver or gold were not just a problem for the French. Through the ages, entrepreneurs① with an eye for a quick buck② have made money for nothing, shaving small fractions off coins to smelt into new currency. "Clipped coins③" introduced new cash into economies, but as well as being illegal, they caused inflation④, leaving people feeling poorer. So when, in 1686, England faced a currency crisis and was carrying a huge national debt, the king, William III, vowed to solve the coinage scandal once and for all. As well as minting new coins, he would plug the hole in the country's finances with a fair way of raising new revenues – a tax on light and air.

England was in dire⑤ need of money. The king's two-pronged⑥ tax-and-mint strategy would restore confidence in an economy burdened with so many clipped coins no one really knew how much silver they had in their pocket. But issuing new currency would be costly. Coins had to contain the correct amount of metal, which meant buying silver and incurring⑦ more debt – and cash was something William didn't have. Even if he did, he wasn't sure that it was worth the value stamped⑧ on it.

光和空气税

　　硬币中的金银变少不仅对法国来说是个问题。多年以来，想发不义之财的企业家一直在空手套白狼，他们从硬币上刮取一小部分铸成新钱。"削边货币"给经济注入新的现金，但这是非法的，这种做法引发了通货膨胀，让人们感到越来越穷。1686年，英国面临货币危机，还持有大量国债，国王威廉三世发誓要彻底解决造币丑闻。除了铸造新币，他还要通过公平的方式增加收入来填补国家的财政缺口——征收光和空气税。

　　英国迫切需要钱。如此多的削边货币给经济带来了沉重负担，没人真正知道自己口袋里到底有多少银子。国王这种税收加铸币的双管齐下政策会重塑信心，但发行新的货币代价很高。硬币要含有定量的金属，这就意味着要买白银，要背负更多的债务，而威廉恰恰没有现金——即使有，也无法保证真的值那么多钱。

　　如果说铸造新币解决削边货币危机相对来说比较简单，那么如何向顽强反抗的国民收税就没那么简单了。被

注释

① recalcitrant [rɪˈkælsɪtrənt] *adj.* 倔强对抗的，拒不服从的

② repeal [rɪˈpiːl] *v.* 撤销，废除

③ assessor [əˈsesə(r)] *n.* 技术顾问，评税员，估价员

④ levy [ˈlevi] *n.* 征兵，征税

⑤ procreation [ˌprəʊkrɪˈeɪʃn] *n.* 生产，生殖

⑥ skirmish [ˈskɜːmɪʃ] *n.* 小规模战斗，小争论，小冲突

⑦ whilst [waɪlst] *conj.* 同时，有时

⑧ outlandish [aʊtˈlændɪʃ] *adj.* 古怪的，奇异的

⑨ intrusive [ɪnˈtruːsɪv] *adj.* 闯入的，打扰的，侵入的

⑩ opprobrium [əˈprəʊbriəm] *n.* 耻辱，羞辱

⑪ bureaucratic [ˌbjʊərəˈkrætɪk] *adj.* 官僚的，官僚主义的

⑫ dwelling [ˈdwelɪŋ] *n.* 住处，处所

⑬ eminently [ˈemɪnəntli] *adv.* 突出地，显著地

If minting new coins to solve the clipped coin crisis was relatively simple, how to tax a recalcitrant① nation was less so. A widely despised hearth tax, in force for a little over twenty years, had only recently been repealed②. If an Englishman's home is his castle, people said that tax assessors③ – who were known as 'chimney men' for this purpose – had no right trooping in to tot up hearths, stoves and chimneys. Older taxes and foreign levies④ had their attractions. A former tax on bachelors, for instance, promoted marriage and procreation⑤ when more people were needed to fight wars, but the king would wait till 1695 when skirmishes⑥ with the French were hotting-up again before inflicting that one on England. A beard tax carried a certain perception of even-handedness, as Henry VIII, who introduced it, wore one himself, but this was not on the cards for William. Increasing the tax on playing cards would raise insufficient revenues; whilst⑦ income tax was completely out of the question. When William considered it, the collective nation almost choked on its breakfast. The very idea that the government could pry into what people earned was an outlandish⑧ invasion of privacy. So William looked out of his palace and hatched a plan. A window tax, though as intrusive⑨ as the hearth tax before it, would be accepted with marginally less opprobrium⑩, he thought.

Casting aside the usual complaints – it would cost too much to collect; it was unnecessarily bureaucratic⑪; there were too many loopholes – William's 'Duty on Lights and Windows' became law in 1696. All houses, except those of the very poorest, paid 2*s* a year, regardless of the number of windows. Dwellings⑫ with between ten and twenty windows paid 6*s*, and anyone with more than twenty windows paid a total of 10*s*. If this seemed clear enough, the incomprehensibly complex rules were eminently⑬ breakable.

大众轻视的壁炉税，强制实施了20多年，最后不得不废止了。如果一个英国人的房子是他的城堡，那么人们说估税员（因估税而被称为烟囱人）无权进来统计壁炉、炉灶和烟囱。旧的税收和涉外税收自有吸引力。例如，以前对单身汉征税，在需要更多人打仗时，促进了婚育，但国王一直等到1695年和法国的冲突再次激烈时，才征收这种税。胡须税让人感到有一定的公平性，因为提出征收此税的人——亨利三世自己就留有胡子，但这种情况绝不可能发生在威廉身上。提高纸牌税可能增加原本不足的税收收入，而所得税根本不可能。当威廉考虑征收所得税时，全体国民气得几乎被早饭噎死。这个想法可以让政府窥探人民的收入，显然严重侵犯了人民的隐私。于是威廉从宫里放眼望去，一计又上心头。他想，虽然窗户税和之前的壁炉税一样扰民，但人们在接受的同时肯定会少骂两句。

　　顶着惯常的抱怨——征税代价太高、没必要如此官僚、有太多漏洞——威廉的"光税与窗税"于1696年成为法律。除了极度贫困的人之外，所有房屋不论有多少窗户，每年固定交2先令；有10~20扇窗户的房屋再交6先令；超过20扇窗户的房屋再交10先令。如果这条规定看起来足够清楚，那么那些复杂到不可思议的规则就非常容易被打破。

▲ 窗税于1696年引入英国，导致许多房主把已有的窗孔堵起来。

The window tax, introduced into England in 1696, meant that many owners bricked up existing apertures.

注释

① exempt [ɪg'zempt] *adj.* 被免除的，被豁免的

② pantry ['pæntri] *n.* 餐具室，食品储存室

③ brick [brɪk] *v.* 用砖建造、砌或铺

④ extravagance [ɪk'strævəgəns] *n.* 奢侈，挥霍，奢侈品

⑤ conjecture [kən'dʒektʃə(r)] *n.* 推测，猜测

⑥ aperture ['æpətʃə(r)] *n.* 孔，洞

⑦ homeowner ['həʊməʊnə(r)] *n.* 自己拥有住房者，房主

⑧ perforate ['pɜːfəreɪt] *v.* 在……上穿孔或打眼

⑨ magistrate ['mædʒɪstreɪt] *n.* 地方法官，治安官

⑩ thwart [θwɔːt] *v.* 阻挠，使受挫折，挫败

Tax evasion began almost at once. Homeowners took to bricking up or painting over windows. Some rooms were exempt①, leading people to dedicate the most improbable of spaces as pantries② or grain stores. As assessments were infrequent, windows could be bricked③ up one day and un-bricked the next. Bribery was commonplace, especially as the local JP, usually a well-connected man of property known to the wealthier landowners, acted as the assessor and would often turn a blind eye to the glazing extravagances④ of influential neighbours. Even the definition of what was or was not a window was a troubling matter of conjecture⑤ that could go either in the landowner's favour or the taxman's. According to contemporary dictionaries, the word derived from 'wind door', signifying 'any aperture⑥ in a building by which light and air are intro-mitted'. Tax inspectors adored this broad definition as it covered just about every hole in the wall. If they didn't like the homeowner⑦ much, it could be costly, as one unfortunate gentleman with a severe damp problem discovered. On the advice of a sanitary expert, a Mr Williams installed four perforated⑧ zinc plates to improve ventilation in his home, only for the taxman to assess it as a new window and place him in a higher tax bracket. When he appealed, magistrates⑨ decided that both parties were wrong. Four zinc panels created four new windows, they said, and they increased Mr Williams' tax further still.

Such cases were widespread until a revised law clarified the position in favour, quite naturally, of the taxman. Individual panes cast in one frame were classed as separate windows, even if they didn't have glass. And even a hole could be a window. In one case, a space in a wall used for shovelling coal was taxed. The taxman nearly always wins.

But not quite. The inventiveness of taxpayers to thwart⑩

几乎马上就开始出现逃税现象。房主们用砖封住窗户或在窗户上涂漆。有些房间是免税的，于是人们又尽力腾出空间作为食品储藏室或粮仓。由于不经常评估，今天拿砖堵住的窗户明天就可以把砖拆掉。贿赂成风，尤其是地方执法官，他们往往人脉很广，认识许多富有人家，他们作为估税员，经常对那些有钱有势的邻居睁一只眼闭一只眼。甚至对窗户的界定也是件麻烦事，房主和收税员各执一词。根据当代词典来看，"窗"这个词由"风"和"门"演变而来，意思是"建筑上让光和空气穿过的洞"。税务稽查员比较喜欢这个广泛的定义，因为它包括了墙上所有的洞。一位不幸家里有严重潮湿问题的先生发现，如果税务稽查员不喜欢房主，那要交的税就多了。这位姓威廉斯的先生听了卫生专家的建议，装了四块有排气孔的锌板，用于促进室内空气流通，收税员把这判定为新窗户，把他列入多交税的范围。他提出申诉，仲裁官判定双方都有错。仲裁官们说四块锌板就造了四扇新窗户，于是他们进一步增加了威廉斯先生的税。

直到修订法案澄清了究竟赞同哪一方的立场之前，这些案例非常普遍。法案自然是赞同收税员的立场了。窗框内的单独窗格被归类为单独的窗户，哪怕没装玻璃也算，甚至一个洞也可以被看作一扇窗户。有一个例子，墙上用于加煤的空隙也被征税了。收税员几乎总是赢家。

但也不完全如此。纳税人总是创造性地反抗政府，到了18世纪20年代，税收下降了。1729年的财政检查发现，"纳税人堵住窗户，地方执法官阻挠评税工作，检查员懒惰无能，真是令人遗憾"。于是法律更加严格，一旦发

注释

① obstruct [əb'strʌkt] vt. 阻塞，堵塞，阻碍

② stiff [stɪf] adj. 不易弯曲、移动、变形的，生硬的

③ garret ['gærət] n. 顶楼，阁楼

④ cellar ['selə(r)] n. 地下室，地窖

⑤ scullery ['skʌləri] n. （炊具食品等的）洗涤室

⑥ bakehouse ['beɪkhaʊs] n. 面包（糕饼）烘房

⑦ brewhouse ['bruːhaʊs] n. 酿酒室，啤酒厂

⑧ exempt [ɪg'zempt] v. 使免除，豁免

⑨ cider ['saɪdə(r)] n. 苹果汁

⑩ tenement ['tenəmənt] n. 房屋，住户

⑪ compensate ['kɒmpenseɪt] v. 补偿，赔偿

⑫ squalor ['skwɒlə(r)] n. 污秽，肮脏，悲惨，贫困

the authorities took such hold that by the 1720s revenues were in decline. A treasury review in 1729 found 'a sorry story of windows stopped up, of JPs obstructing① the work of assessment, of surveyors lazy and incompetent'. The law was tightened again, with stiffer② penalties for anyone caught blocking and reopening windows. The tax now covered:

skylights, windows, or lights, however constructed, in staircases, garrets③, cellars④, passages, and all other parts of the house, to what use or purposed soever, and every window or light in any kitchen, scullery⑤, buttery, pantry, larder, washhouse, laundry, bakehouse⑥, brewhouse⑦ and lodging room occupied with the house, whether contiguous to or disjoined from the dwelling-house.

Construction of outhouses, an unintended consequence of window tax and from which they were exempt⑧, slowed.

With the country needing more money again, revenues started to rise. William III was long gone, but by this time the Seven Years War in Europe needed funding and other taxes like the cider⑨ levy and one that funded the war in America had been repealed. In 1766, window tax was extended again, bringing in homes with just seven windows. The result changed housing and architecture in Britain at all points on the social scale. Although window tax had originally been designed as a progressive tax that hit the rich more than the poor, because windows in whole buildings were assessed, not those in individual garrets or flats, landlords of tenement⑩ houses found themselves heavily taxed. To compensate⑪, they upped rents on existing properties or boarded windows up, leaving tenants now not just in squalor⑫, but in dark, airless squalor to boot. In Edinburgh, entire rows of houses were built without a single window in the bedrooms.

现任何人堵塞或重开窗户，都要严惩不贷。这时的窗税
包括：

> 楼梯、阁楼、地窖、走廊及房子其他部分
> 内不论何种结构、何种用途的天窗、窗户或透光
> 口；房主占用的厨房、碗碟储藏室、酒储藏室、
> 食品储藏室、肉储藏室、洗涤室、洗衣房、面包
> 烘房、酿酒室和出租房（无论与住所连接与否）
> 的窗户或透光口。

于是，外屋（本来是免收窗税的）的修建也减缓了。

由于国家需要更多的钱，税收又开始提高。威廉三
世早就去世了，但此时，欧洲的英法七年战争需要资金，
而其他税收，例如苹果酒税和为美国战争提供资金的一项
税收已经被废止了。1766年，窗税再次扩大，把只有7扇
窗户的房屋也纳入征税范围。结果从社会层面上改变了英
国的住房和建筑。尽管窗税最初设计为累进税，富人比穷
人交的税多，但由于评税是针对整栋楼的窗户，而不是针
对单个阁楼或公寓的窗户，结果房东发现自己缴的税比较
多。为了弥补亏空，他们要么提高租金，要么用木板把窗
户围起来，使房客住的地方不仅肮脏，而且黑暗、不通风
加肮脏。在爱丁堡，整排的房子卧室都没修一扇窗户。在
其他地方，卧室的窗户被封或涂上油漆，可以想象在那里
度过的时光是多么黑暗。相反，拥有众多房产的富家大族
则找到了新的炫富方式。设计师给富丽奢华的乡间住宅、
宫殿和通道装饰了尽可能多的窗户，只要能引起街坊邻居

注释

① gentry ['dʒentri] *n.* 绅士们，贵族们

② opulent ['ɒpjələnt] *adj.* 富裕的，豪华的

③ ostentatious [ˌɒsten'teɪʃəs] *adj.* 好夸耀的，炫耀的

④ sabre-rattling ['seɪbə 'rætlɪŋ] *n.* 武力威胁，炫耀武力

⑤ guinea ['ɡɪni] *n.* 畿尼（英国的旧金币，值1镑1先令）

⑥ hue [hju:] *n.* 色彩，色调

⑦ evade [ɪ'veɪd] *v.* 逃避，躲避 tax-evading 逃税

⑧ execution [ˌeksɪ'kju:ʃn] *n.* 依法处决，实行，执行

⑨ galling ['ɡɔ:lɪŋ] *adj.* 使烦恼的，难堪的

⑩ consternation [ˌkɒnstə'neɪʃn] *n.* 惊愕，惊慌失措，惊恐

⑪ whoop [wu:p] *v.* 叫喊，喘息，高声说

Elsewhere, bedroom windows were blocked or painted on the logical assumption that the hours spent there were dark anyway. By contrast, at the top end of the scale, the landed gentry① found a new way of demonstrating their wealth. Designers of opulent②, ostentatious③ country homes, palaces and avenues packed in as many windows as they could. The master would willingly pay the tax as long as the neighbours noticed.

Still the government wanted more. In the 1780s, with the French sabre-rattling④ and Britain on a war footing, taxes rose again. Fortunately for the government, the prime minister, William Pitt the Younger, had a talent for taxes. A guinea⑤ imposed on the white powder gentlemen used to give their wigs an off-white hue⑥ became the wig tax, but raised an insignificant amount and upset the gentry. Hat tax similarly targeted the rich who, if they wanted to get ahead, got a hat – or rather several hats, one for every occasion. Tax-evading⑦ hat wearers and milliners faced the threat of execution⑧. Each piece of men's headwear had to have the Revenue's stamp on the inside. Forging a stamp, whether on hats, wig-powder packaging, playing cards, or indeed on anything taxable, could, and sometimes did, lead to the gallows.

Pitt's initiatives didn't stop with headwear. A tax on sporting dogs came in 1786. A brick tax at 5*s* per 1,000 bricks raised yet more money (but also caused manufacturers to make buildings from larger bricks). A levy on 'watches and persons in possession of clocks' taxed time. Even soap, a luxury in the Georgianera, came with a contribution to government coffers. But the most galling⑨ tax of all was income tax – the very charge that even William III couldn't bring himself to introduce all those years ago. Income tax caused untold consternation⑩ among the landed voting classes and, when introduced in 1799, didn't last long. In 1815, to whoops⑪ of joy from MPs – the very

的注意，主人肯定乐意交税。

政府还是觉得不够。18世纪80年代，由于法国不断武力挑衅，英国实施了战时体制，税收再次提高。政府真幸运，首相小威廉·皮特极具收税天赋。对男士用于把假发染成白色的白粉征收1畿尼（英国的旧金币，合1镑1先令），称为假发税，但收入不多，还得罪了上流社会。帽子税同样针对富人，如果他们想获得成功，就会买一顶帽子或几顶帽子，以应对不同的场合。偷税的戴帽者和帽子制造贩卖商都可能被处以死刑。男人的头饰内侧都要有已交税的印章。伪装印章，在帽子、假发粉包装、纸牌或所有要交税的东西上伪造印章都可能会被判绞刑，有时候确实会被绞死。

皮特的举措并未止于头饰。1786年开始对猎犬征税。砖税（每1000块砖收5先令）使税收收入增加，不过也导致建筑商用更大块的砖盖楼。手表和钟表所有者税是对时间征的税。甚至连乔治王朝时期的奢侈品肥皂也为政府金库贡献了税收。不过最恼人的税要数所得税，这项税收连多年前的威廉三世都不敢推行。所得税使有土地、有投票权的人们无比恐慌，1799年开始实施后没持续多久。1815年，让议员们——交税最多的人群——欢呼雀跃的是所得税被废止了，而且所有相关文件全部作废。直到1842年才又出现了所得税，即使那时也和现在一样，是暂时税，只会征收一年（但每次预算中都会恢复）。

所得税被取消了，那就再次提高窗税来弥补财政赤字。这次遭到了更为严重的鄙视，于是税务员在评估房屋时越来越需要警察保护，甚至连他们自己也开始鄙视这

注释

① deficit ['defɪsɪt] n. 不足额，赤字，亏空

② despise [dɪ'spaɪz] v. 鄙视，看不起

③ hazard ['hæzəd] n. 危险，冒险的事

④ unsanitary [ʌn'sænətri] adj. 不卫生的，有碍健康的

⑤ ventilation [ˌventɪ'leɪʃn] n. 空气流通，通风设备

⑥ amid [ə'mɪd] prep.（表示位置）在……中间，（表示环境）处于……环境中

⑦ scathing ['skeɪðɪŋ] adj.（言辞、文章）严厉的，不留情的

⑧ indisputable [ˌɪndɪ'spju:təbl] adj. 无可争辩的，不容置疑的

⑨ descend [dɪ'send] v. 下来，下降

people paying most – it was repealed and all documentation relating to it pulped. Income tax wouldn't be seen again until 1842 and even then, like now, it was a 'temporary tax' that would only last a year (but is renewed at each Budget).

With income tax gone, window tax rates again rose to cover the deficit①. But it was despised② all the more and tax inspectors increasingly sought police protection when assessing homes. Even they began to despise it. At one end, they were taxing their important neighbours; at the other they were contributing to increasingly evident health hazards③ caused by lack of light and air. In greater numbers, working people lived in damp, unsanitary④ conditions with inadequate ventilation⑤. Much of the trouble was ascribed to window tax. Amid⑥ worsening conditions, respected architect Henry Roberts reported on the link between window tax and poor health. His findings were scathing⑦; the link indisputable⑧. In the new Victorian age with the world about to descend⑨ on the 1851 Great Exhibition, this was a time to clean up. Just before the exhibition opened, window tax was repealed, although a tax on glass remained for six more years. New ways other than window tax would have to be found to fund government. And they have yet to run out of ideas.

项税收了。他们一边对有权有势的邻居收税，一边为缺乏日照和空气流通导致的日益明显的健康危险作着贡献。居住在潮湿、肮脏且没有足够空气流通环境中的工人越来越多，问题主要在于窗税。令人尊敬的建筑师亨利·罗伯茨（Henry Roberts）身处日益恶化的环境之中，报道了窗税与健康不良之间的联系。他的见解十分犀利：两者之间的联系无可争辩。在新的维多利亚时期，全世界的人即将涌向1851年的万国工业博览会，该大赚一笔了。博览会开幕前不久，窗税被废止了，但对玻璃的税收依旧持续了六年。应该找个除了窗税之外的新方法来资助政府。他们还没有思路枯竭，有的是办法。

PART TWO

第二部分

THE INTERNATIONAL
'HOT AIR' AIRLINE

It was to be the first international airline and the first airmail service. But the difficulties of launching the Aerial① Transit② Company, which planned to shrink③ the globe from its headquarters④ at a lace factory in Somerset, proved insurmountable⑤. This was 1842, the age of steam, and getting an aircraft off the earth's surface was proving to be a lot of hot air.

Unlike the story of Icarus, who attempted to flee⑥ Crete⑦ by flapping feathered wings designed by his dad Daedalus, the international steam-powered airline is no myth⑧. Inventors and entrepreneurs William Samuel Henson and John Stringfellow followed in a succession of aspiring⑨ aeronautical⑩ engineers who, since Icarus, have tried to defy⑪ gravity with an array⑫ of hapless⑬ devices. In 1540 in Portugal, João Torto believed he'd bettered Daedalus' design by using two pairs of cloth wings while wearing a helmet in the shape of an eagle's head, but this only served to double the speed with which he plummeted⑭ to earth after jumping from a cathedral tower. In 1712 Frenchman Charles Allard strapped⑮ on wings of an improved design and launched himself from the Terrasse de St Germain in the

国际"蒸汽"航线

　　国际"蒸汽"航线差点成为最早的国际航线和最早的空邮服务，但成立航空运输公司的重重困难却使其半路夭折。航空运输公司原本计划从位于萨默塞特的一家花边厂的总部发展航空事业，以期缩短世界各地之间的距离。时值1842年，正是蒸汽机的时代，而使飞行器脱离地面似乎需要大量蒸汽。

　　在古希腊神话中，伊卡洛斯试图利用父亲代达罗斯设计的翅膀逃离克里特岛，但国际蒸汽动力航线却不是神话。在**伊卡洛斯**之后，发明家、企业家威廉·萨缪尔·亨森（William Samuel Henson）和约翰·斯特林费罗（John Stringfellow）之前，许多雄心勃勃的航空工程师试图利用一些飞行器脱离地心引力，但不幸均以失败而告终。1540年，葡萄牙人若昂·拓特（João Torto）认为，通过两对布翅膀和一项鹰式头盔可以改进代达罗斯的设计，但这反而成倍加快了他从大教堂塔顶垂直落地的速度。1712年，法国人查尔斯·阿拉德（Charles Allard）身背改进的翅膀从

注释

伊卡洛斯是希腊神话中代达罗斯的儿子，与代达罗斯使用蜡和羽毛造的翼逃离克里特岛时，他因飞得太高，双翼上的蜡被太阳融化跌落水中丧生，被埋葬在一个海岛上。

注释

① steeplejack ['stiːpldʒæk] n. 高空作业人员

② tightrope ['taɪtrəʊp] n. 绷紧的绳索，绷紧的钢丝

③ taut [tɔːt] adj. 拉紧的，绷紧的

④ cord [kɔːd] n.（结实的）粗线，细绳，一根粗线（或细绳）

⑤ extravagant [ɪkˈstrævəgənt] adj. 奢侈的，挥霍的，铺张浪费的

⑥ abseil ['æbseɪl] v. 绕绳下降（用绳缠绕着身体，双脚蹬陡坡或峭壁自己放绳下滑）

⑦ spire ['spaɪə(r)] n.（教堂等顶部的）尖塔，尖顶

⑧ fatal ['feɪtl] adj. 致命的

⑨ airborne ['eəbɔːn] adj. 升空

⑩ trajectory [trəˈdʒektəri] n.（射体在空中的）轨道，弹道，轨迹

⑪ bobbin ['bɒbɪn] n. 线轴，绕线筒

⑫ glider ['glaɪdə(r)] n. 滑翔机

direction of Bois du Vésinet, but only completed the Terrasse de St Germain part of the journey before dying from multiple injuries. And in 1739 steeplejack① and occasional tightrope② walker Robert Cadman entertained the residents of Shrewsbury with his own spectacular and rather messy death, while attempting to soar from the spire of St Mary's Church to the far bank of the River Severn with the help, or hindrance as it turns out, of an overly taut③ piece of cord④ that snapped under his weight. This may have been more an extravagant⑤ abseil⑥ than genuine flight, but the result was much the same. Cadman's wife, passing round a hat below, dropped all the donations when told how he had been 'dashed to pieces' while her back was turned. A stone memorial now commemorates the achievement, recalling Cadman's:

> attempt to fly from this high spire⑦
> across the Sabrine he did acquire
> His fatal⑧ end.

So the fact that no one had actually been airborne⑨ for more than a few seconds – and only then in a strictly vertical and downwards trajectory⑩ at speeds they hadn't expected – should have made the task of starting the world's first international airline rather daunting. But in the industrial age, along with the rapid advancements of scientific theory, inventions were coming thick and fast. There was no reason that man shouldn't get into the air and return to earth thousands of miles from his starting point with his limbs attached in the traditional formation. Indeed Henson, the son of a lace factory owner, and Stringfellow, a toolmaker who made bobbins⑪, thought they knew how.

With the expertise of their friend Sir George Cayley, who had designed the first glider⑫ to carry a human being and who

圣日耳曼平台朝维西内森林的方向飞去，但刚飞出圣日耳曼平台，就多处受伤而死。1739年，高空作业工人及兼职钢丝演员罗伯特·卡德曼（Robert Cadman）以其公开演出甚至玩儿命来逗什鲁斯伯里的居民开心。当时，他试图借助系在身上的一根绳子从圣玛丽教堂的尖塔顶飞到塞文河对岸，但这根绳子最后却成了羁绊。这可能更像是荒诞的沿绳滑下，而不像真正的飞行，但结果却一样。卡德曼的妻子当时正端着帽子四处收取小费，当她回过头来，人们告诉她卡德曼是如何"摔得粉身碎骨"时，她吓呆了，把钱撒了一地。今天在这里树了一座石碑，以纪念卡德曼的成就，上面写道：

> 他试图飞下高塔，
> 飞越塞文河，
> 但不幸在此殒命。

所以没人曾经在空中真正飞行超过几秒（就这几秒也是在以难以预料的速度垂直向下降落）的事实使开创世界上首条国际航线着实令人望而生畏。但在工业时代，随着科学理论的迅速发展，发明成果大量涌现，人类如果再不能以传统方式把四肢固定，进入天空，然后再从千里之外落回地面，似乎就没有理由了。事实上，花边厂厂长的儿子亨森和制线筒的模具工斯特林费罗认为他们知道该怎么办了。

在朋友乔治·凯利（George Cayley，乔治·凯利爵士设计了第一架载人滑翔机，有时被称为"航空学之父"）

注释

① self-propelled [self prə'peld] *adj.* 自力推进的

② heyday ['heɪdeɪ] *n.* 最为强大（或成功、繁荣）的时期

③ taxidermy ['tæksɪdɜːmi] *n.* 动物标本剥制术（将动物充填以支撑物，以表现出其生前外形）

④ paddle ['pædl] *n.* 船桨

⑤ stoked [stəʊkt] *adj.* 满足的

⑥ deft [deft] *adj.* 熟练的，灵巧的，机敏的

⑦ monster ['mɒnstə(r)] *n.* （传说中的）怪物，怪兽

⑧ prevailing [prɪ'veɪlɪŋ] *adj.* 普遍的，盛行的，流行的

⑨ monoplane ['mɒnəpleɪn] *n.* 单翼飞机

⑩ wingspan ['wɪŋspæn] *n.* 翼展，翼幅

⑪ practicality [ˌpræktɪ'kælɪtɪ] *n.* 可行性，适用性

is sometimes described as the father of aviation, they calculated what it would take to build a passenger-carrying, self-propelled[①] aircraft. Having studied birds in flight extensively and – this being the heyday[②] of natural history – practiced their trajectory across a room with stuffed ones, they felt they were comfortable with the concepts of movement in three dimensions. 'My invention,' said Henson in his 1842 patent, 'will have the same relation to the general machine which the extended wings of a bird have to the body when a bird is skimming in the air.' Such aeronautical taxidermy[③] was all very well, but they also needed mechanics. A boiler, a paddle[④] wheel, and somewhere to sit while the pilot stoked[⑤] the engine would suit perfectly.

Before beginning international flight operations, they first had to build an aircraft. Henson's designs were elegant and Stringfellow's lightweight engines, in deft[⑥] contrast to the other industrial monsters[⑦] of the age, inspired. Under prevailing[⑧] UK patent rules that didn't require evidence that inventions worked, they would create a machine 'to convey letters, goods and passengers from place to place through the air.' They called it the Aeriel or the Aeriel Steam Carriage; a monoplane[⑨] with a wingspan[⑩] of 150ft that would carry a dozen passengers 1,000 miles, although with a top speed of just 50 mph, this would mean 20 airborne hours, which was asking a lot for a lightweight engine of 50hp (37 kw). In terms of practicalities[⑪], 2 square feet of supporting surface added a pound of weight, requiring the engine to generate 20hp per ton to stay airborne. It worked on paper.

With a patent granted and investment raised from a small group of friends, attention turned to the second key aspect of the plan: publicity. In this, Henson and Stringfellow went to town. If you're going to launch the world's first international airline, then the world has to know about it. To Sir George

爵士的专业知识的帮助下，他们想出了用什么来制造一架载人自行飞机。当时是博物学的全盛期，他们广泛研究了鸟类飞行，又在室内研究了鸟类剥制标本的飞行轨迹，自认为对三维运动的概念十分了解了。"我的发明，"亨森在1842年申请专利时说，"相对于整个机器的作用就像鸟在天空飞行时伸展开的翅膀相对于身体的作用是一样的。"这种可以飞行的鸟类标本模型确实不错，但它们还需要机械的部分。若再配上蒸汽机、叶轮和飞行员座椅就臻于完美了。

　　在开始国际飞行业务之前，他们先得制造一架飞机。亨森的设计很精致，而斯特林费罗的轻型发动机则与那个时代的其他工业怪物不同，创意极佳。当时的英国专利法不需要证明发明起作用，据此，他们要创造一台"在空中运输信件、货物和乘客"的机器。他们称为"空中马车"或"空中蒸汽马车"；一架翼展达45.72米（150英尺）的单翼飞机可以把数十名乘客送到1609千米（1000英里）之外，即便最高时速（仅为80千米/时，即50英里/时），也要在空中飞行20个小时，这对于一台37千瓦（50马力）的轻型发动机来说负担不轻。在实践中，每平方米翼面增加2.4千克（每2平方英尺翼面增加1磅）的重量，需要发动机达到每吨15千瓦（20马力）才能在空中飞行。现在只是纸上谈兵。

　　专利获批之后，他们从一些朋友那里筹集了部分资金，但这一计划又面临着另一个关键问题：宣传。为了宣传，亨森和斯特林费罗去了城里。如果你想开创世界上首条国际航线，就得让世人了解它。乔治·凯利爵士想逐步

注释

① consternation [kɒnstə'neɪʃ(ə)n] n. 惊愕，惊恐

② flyer ['flaɪə] n. 传单

③ airfare ['eəfeə(r)] n. 机票费用，飞机票价

④ gear [gɪə] n. 排挡，齿轮，传动装置

⑤ contra-rotate [ˌkɒntrərəʊ'teɪt] v. 逆时针旋转

⑥ undaunted [ʌn'dɔːntɪd] adj. 勇敢的，无畏的

⑦ tray [treɪ] n. 托盘，文件盒

⑧ tapestry ['tæpɪstrɪ] n. 织锦，挂毯，绣帷

⑨ rum [rʌm] adj. 古怪的，奇特的

⑩ soldier ['səʊldʒə] v. 坚持干

⑪ aeronautical [ˌeərə'nɔːtɪkl] adj. 航空的，航空学的

⑫ redundant [rɪ'dʌnd(ə)nt] adj. 多余的，过剩的

⑬ lace [leɪs] n. 花边，饰带

⑭ mill [mɪl] n. 工厂

Cayley's consternation① – he was to withdraw eventually from investment – posters and flyers② began to appear of the plane in the unlikeliest locations: here's the Aeriel in flight over London; and here it is again, this time over the Giza pyramids; and now India. An advertisement of the plane soaring over China was a particular favourite of the two inventors, who hoped to build eastern markets for British commerce.

China, though, would have to wait, for Chard had to be conquered first. Residents of the Somerset town were to be Henson and Stringfelllow's first audience – and, it was to be hoped, future airfare③ paying customers. The Ariel might not have been able to carry passengers at this – nor indeed any – stage, but the prototype was impressive enough: the first plane of modern construction with a 12ft wingspan, three-wheeled landing gear④ and power from two contra-rotating⑤ bladed propellers. Its first run, in contrast, was a thunderous disappointment all round. After shuddering down a short ramp, instead of taking to the air, it came to a pitiful stop.

Undaunted⑥, Henson and Stringfellow set to work on a bigger model, with a broader 20ft wingspan and a steam engine that delivered greater power. Marketing was also stepped up. Promotional handkerchiefs, trays⑦, wall tapestries⑧ and lace placemats joined the posters and newspaper advertisements. It was nothing but hot air. Over the course of three years from 1844, the larger model plane was tested over and over and over again. It never flew. Finally, at his wit's end, Henson took drastic action; he married his girlfriend and, packing up the whole enterprise as a rum⑨ job, immigrated to the States, presumably by steamer. Stringfellow soldiered⑩ on, his aeronautical⑪ ambitions higher than ever. And for him at least, success followed.

In a redundant⑫ lace⑬ mill⑭ in June 1848, the model plane

收回投资，海报上的宣传让他大为惊愕，飞机出现在了最不可能出现的地方：空中蒸汽马车飞过伦敦，飞过**吉萨金字塔群**，飞过印度。而一份飞机飞过中国的广告体现了两位发明人的特别兴趣，他们希望为英国贸易开拓东方市场。

不过中国市场还得等一等，因为先得征服英国。萨默塞特镇的居民成了亨森和斯特林费罗的首批观众，当然他俩也希望他们成为未来的机票购买客户。空中蒸汽马车在当时乃至今日可能无法载客，但样机实在惊人：这架现代制造的首款飞机翼展达3.66米（12英尺），配有三轮起落架，以倒转双叶螺旋桨为动力。然而，它的首次试飞却令所有人大跌眼镜。它在一条短跑道上一阵颤动之后，并没有飞向天空，反而可怜地停了下来。

但是亨森和斯特林费罗毫不气馁，又开始制造另一架更大的样机。这架样机翼展达6.10米（20英尺），发动机功率更大。营销活动也在紧锣密鼓地进行。促销手帕、托盘、墙上挂毯、花边餐具垫等配合海报、报纸广告，搞得如火如荼，但也无非是乱吹一通。1844年之后的三年时间里，他们又反复试验了这一更大的模型飞机，但还是没有飞起来。最后亨森实在无计可施了，就毅然和女朋友结婚，扔下公司，举家迁到了美国（想必是乘坐轮船去的）。斯特林费罗则更加士气高昂地从事着自己的航空壮举，最后他终于成功了。

1848年6月，在废弃的花边厂里，这架模型飞机起飞了。斯特林费罗的飞机飞起来了！飞行时间虽然很短暂，但总算是飞起来了。据一些报道称，这架飞机直飞了大约

注释

吉萨金字塔群位于尼罗河三角洲的吉萨，是由第四王朝的3位皇帝胡夫、哈弗拉和门卡乌拉在公元前2600—公元前2500年建造的。

注释

① accolades [akɔɪadz] n. 赞美，表扬（accolade 的复数形式）
② horsepower ['hɔːspaʊə] n. 马力（功率单位）
③ navigation [nævɪ'geɪʃ(ə)n] n. 航行，航海
④ combustion [kəm'bʌstʃ(ə)n] n. 燃烧，氧化，骚动
⑤ overtake [əʊvə'teɪk] vt. 超过，赶上

was launched from an inclined wire. Stringfellow had flight! Short flight, but flight nevertheless. According to some reports, the plane flew straight for about 30ft (although other reports dismiss its achievements as nothing more than 'a short hop'). Delighted whatever the distance, Stringfellow repeated the exercise and said the plane even sometimes gained altitude – which would be handy for the journey to China. Accolades① followed. When the Aeronautical Society awarded Stringfellow a £100 prize, Scientific American magazine was impressed. He was, they claimed, responsible for 'probably the lightest ever steam engine ever constructed'.

By 1869, Stringfellow had developed the engine further. It now turned 3,000 revolutions a minute and, reported Scientific American, just 'three minutes after lighting the fire the pressure was up to 30 pounds and in seven minutes the full working pressure of 100 pounds, generating just over one horsepower②'. It was laughably insufficient to get passengers to the pyramids or New Zealand. It wouldn't even get a full size plane airborne at all. But Henson and Stringfellow's invention did become the first powered plane in the world to fly. While only a model, it was to form a landmark on the way to the first passenger flights. Just before his death in 1883, Stringfellow said: 'Somebody must do better than I before we succeed with aerial navigations③.' Just twenty years later, with the internal combustion④ engine overtaking⑤ steam as the power of choice, the Wright Brothers were in the air.

9米（30英尺），尽管有些报道蔑称这一成就无非是"短短的一跳"。无论距离长短，斯特林费罗都感到非常高兴，于是反复试验，并称飞机甚至有时能提升飞行高度，他说飞机终有一天会达到可以去中国旅行的水平。荣誉接踵而至。皇家航空学会奖励了斯特林费罗100英镑，他也给《科学美国人》杂志留下了深刻的印象，他们称斯特林费罗对"制造可能是有史以来最轻的蒸汽机"功不可没。

截至1869年，斯特林费罗进一步改进了蒸汽机。《科学美国人》报道称："现在这台蒸汽机1分钟可以转3000转，压力在点火3分钟之后就能达到13.61千克（30磅），7分钟之内达到45.36千克（100磅）的工作压力，产生的推力大于0.74千瓦（1马力）。"遗憾的是，它仍然不能把乘客送到金字塔或新西兰，甚至不能让一架真实大小的飞机起飞。不过亨森和斯特林费罗的发明确实成为世界上首架飞行的动力飞机。尽管只是一架模型飞机，但却是人类载人飞行之路上的一座里程碑。斯特林费罗于1883年去世，他临终时说："在我们的空中飞行成功之前，一定会有人做得比我好。"短短20年之后，随着内燃机代替蒸汽机成为最佳动力选择，莱特兄弟飞向了蓝天。

ABANDONED

THE 'SPRUCE GOOSE'

注释

① eccentric [ɪk'sentrɪk; ek-] *adj.*
古怪的，反常的
② maverick ['mæv(ə)rɪk] *adj.* 行
为不合常规的，特立独行的
③ cantilevered ['kæntlivəd] *adj.*
悬臂式的
④ aviator ['eɪvɪeɪtə] *n.* 飞行员
⑤ cog [kɒg] *n.* 钝齿，雄榫
⑥ flop [flɒp] *n.* 失败，砰然落下
⑦ glamour ['glæmə] *n.* 魅力
⑧ audacious [ɔː'deɪʃəs] *adj.* 无畏
的，鲁莽的
⑨ spruce [spruːs] *n.* 云杉

Howard Hughes surely defined the word eccentric①. One of the richest men in the world, he was famous for his maverick② movie making, addiction to drugs and love of beautiful actresses. Stories of the billionaire's bizarre behaviour are legion too. Legend has it that he once gave staff precise instructions on how to lift a toilet seat. On another occasion, he became obsessed with designing a complicated cantilevered③ bra for one of the stars in his movie *The Outlaw*. Hughes was also an accomplished aviator④ and innovative aircraft manufacturer who was, for a time, considered a crucial cog⑤ in visionary plans to defeat Hitler.

The early twentieth century gave birth to many unsuccessful aircraft. There was the Bristol Brabazon, for example, the trans-Atlantic flop⑥ of the 1940s which was just too big, too expensive and too luxurious for the job. But none of the many flying failures of the century had quite the glamour⑦ and audaciousness⑧ of Howard Hughes' H-4 Hercules, better known as the 'Spruce⑨ Goose'. In its day it was the world's largest plane, and even today still holds the record for the longest wingspan of any aircraft. It was an invention that had all the

"云杉鹅号"飞机

霍华德·休斯（Howard Hughes）绝对是个怪人。他是世界上最富有的人之一，因拍摄特立独行的电影、吸毒成瘾和追求漂亮女演员而出名。关于这位富豪的奇闻逸事非常多。传说他曾给职员明确说明如何抬起马桶盖；还有一次，他沉迷于为他的电影《歹徒》中一位女演员设计复杂的悬臂式内衣。休斯同时也是一位成功的飞行员和具有创新精神的飞机制造商，曾经一度被视为击败希特勒计划中的关键人物。

20世纪初制造了许多失败的飞机。例如，20世纪40年代布里斯托飞机公司设计生产的"布拉巴宗"客机就是一次失败，因为跨大西洋航线市场还不足以运营这样一款过于庞大、昂贵、豪华的飞机。然而，在20世纪众多飞行失败的案例中，没有一个像休斯的**H-4大力神**一样大胆创新，充满魅力，人们通常把这架飞机称为"云杉鹅号"。当时，它是世界上最大的飞机，甚至至今仍然保持着最大翼展的纪录。这架飞

注释

H-4大力神中的大力神指赫拉克勒斯（Ηρακλῆς），他是古希腊神话中的伟大英雄，是主神宙斯与阿尔克墨涅之子，因其出身而受到宙斯的妻子赫拉的憎恶。他神勇无比、力大无穷。

注释

① grandeur ['grændʒə] n. 壮丽，
庄严，宏伟
② foible ['fɔɪb(ə)l] n. 弱点，小缺
点，癖好
③ splash [splæʃ] vi. 溅湿，溅开
④ aviation [eɪvɪ'eɪʃ(ə)n] n. 航空，
飞行术，飞机制造业
⑤ trophy ['trəʊfɪ] n. 奖品，战利
品，纪念品
⑥ subsidiary [səb'sɪdɪərɪ] n. 子公
司，辅助者
⑦ embrace [ɪm'breɪs; em-] v. 拥
抱，信奉
⑧ overrule [əʊvə'ruːl] vt. 否决，
统治
⑨ reconnaissance [rɪ'kɒnɪs(ə)ns]
n. [军] 侦察，勘测
⑩ duly ['djuːlɪ] adv. 适当地，充分
地，按时地

grandeur①, eccentricity and capacity for controversy as the man himself.

Despite his famous foibles②, Hughes, born in 1905 to an oil-industry entrepreneur, was an undoubtedly clever man. At 14 he took his first flying lesson and it led to a lifelong love of flight. Later, in 1930, he would splash③ out a then-vast sum of $4 million to make a First World War flying film called *Hell's Angels*. In 1932 he set up his own aviation④ company, Hughes Aircraft, and went on to help design a number of planes, win flying trophies⑤ and set new air speed records. In 1938 he even flew right round the world breaking Charles Lindbergh's New York to Paris record in the process.

When the war came along Hughes' aircraft company was only employing a handful of people. By the end of the Second World War it would be employing tens of thousands, and along with aircraft development his subsidiaries⑥ would go on to supply the US military with everything from radar to air-to-air missiles and on board fire control systems. In 1942 he was approached by the industrialist and shipbuilder Henry Kaiser, who had a plan to tackle the huge losses being suffered by US shipping at the hands of German U-boats. A massive 800,000 tons of shipping had been sunk and Kaiser felt that the way round this was to build a huge aeroplane, capable of carrying 750 troops, and even tanks, across the Atlantic. Hughes enthusiastically embraced⑦ Kaiser's idea and together they managed to obtain $18 million in funds to build a huge flying boat to be called the HK-1. According to later reports, President Roosevelt himself overruled⑧ experts to give Hughes the contract, along with another for a long-range reconnaissance⑨ plane, the XF-11.

The pair were told to build three HK-1 prototypes to be ready within two years, and work duly⑩ began building the

机是一项伟大而怪异的发明，像休斯本人一样充满了争议。

休斯于1905年出生在一个石油企业家庭，尽管有些小缺点，但他无疑是个聪明人。他14岁开始学习飞行，从此飞行成了他一生的爱好。1930年，他想花400万美元拍摄一部以第一次世界大战为题材的空战大片——《地狱天使》，这在当时可不是一笔小数目。1932年，他创办了自己的航空公司——休斯飞机公司，继续设计飞机，赢取飞行奖杯，创造新的飞行速度纪录。1938年，他甚至环球飞行，打破了查尔斯·林德伯格从纽约飞往巴黎的航行纪录。

第二次世界大战爆发时，休斯的飞机公司只雇用了几个人。而当战争结束时，他已经拥有数万名员工，并且随着航空业的发展，他的子公司仍然是美军的供应商，供应货物一应俱全，包括雷达、空对空导弹、机载火控系统等。1942年，实业家及造船专家亨利·凯撒来找他。凯撒计划解决由德国U型潜艇造成的美国航运的巨大损失问题。总计80万吨的航运船被击沉了，凯撒认为解决办法是建造一架大型飞机横跨大西洋，飞机能容纳750名士兵，甚至坦克。休斯非常赞成凯撒的想法，他们共同筹资1800万美元，准备建造一架巨型水上飞机，计划命名为HK-Ⅰ。据后来的报道，罗斯福总统亲自否决了专家的提议，与凯撒签订了订购合同，外加订购一架远程侦察机XF-Ⅱ。

政府要求休斯和凯撒在2年内造3架HK-Ⅰ，这个项目在加利福尼亚州南部按时展开。像休斯很多的

注释

① epic ['epɪk] n. 史诗，叙事诗

② tailfin [teɪl'fɪn] n. 尾翼，垂直翼

③ cylinder ['sɪlɪndə] n. 圆筒，汽缸

④ moniker ['mɒnɪkə] n. 绰号，名字

⑤ birch [bɜːtʃ] n. 桦木，桦树

⑥ plywood ['plaɪwʊd] n. 夹板，胶合板

⑦ debilitating [dɪ'bɪlɪteɪtɪŋ] adj. 使衰弱的

⑧ dockside ['dɒksaɪd] adj. 码头前沿的；n. 坞边

⑨ hanger ['hæŋə] n. 衣架，挂钩

⑩ amid [ə'mɪd] prep. 在其中，在其间

⑪ lewd [luːd] adj. 好色的，粗俗下流的

⑫ allegation [ælɪ'geɪʃ(ə)n] n. 指控

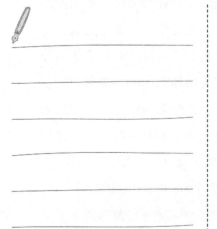

plane on a site in Southern California. Like many of Hughes' films, the final design was truly epic①. The wingspan of the Spruce Goose, as it was nicknamed by a newspaperman of the day, was 320ft. To put that in a modern day context, the Airbus A380's wingspan is a mere 261ft, and a Boeing 747's just 211ft. The wings of the Goose were 13ft thick and the tailfin② alone was the height of an eight-storey building. The beast weighed 400,000lbs and its propellers were more than 27ft long. Together the plane's eight 5,000lb, 28 cylinder③ engines boasted 24,000hp, while the plane could carry 14,000 gallons of fuel. All the more remarkable, because of wartime restrictions, it had to be made out of wood rather than metal – hence its moniker④. In fact the Spruce Goose was mostly made out of birch⑤ plywood⑥, not spruce.

But all the time needed to carry out research on such an enormous plane, as well as supply problems and Hughes' debilitating⑦ perfectionism, meant that construction dragged; and in 1944 Kaiser quit the project, sensing the way the wind was blowing. The end of the war came and went and the aircraft, now officially called the H-4, still hadn't made it to its dockside⑧ hanger⑨. But Hughes didn't care; he shovelled another $7 million of his own money into the plane's manufacture despite the fact that, with the conflict over, the original need for such an aircraft had gone. Then, in summer 1946, Hughes suffered terrible injuries when, piloting his other experimental plane for the government, the XF-11, he crashed into a housing estate. That wasn't his only difficulty. The government had started to ask questions. Where exactly, it wanted to know, had all its money gone? Why had no aircraft been delivered in four years?

In 1947 the US Senate War Investigating Committee began its hearings into Hughes' company, amid⑩ lewd⑪ allegations⑫ of

电影一样，最终的设计也确实如史诗一般。"云杉鹅号"是当时一位记者给这架飞机起的绰号，它的翼展为97.54米（320英尺）。而现在，空中客车A380的翼展仅为79.55米（261英尺），波音747的翼展也才64.31米（211英尺）。而"云杉鹅号"的机翼厚3.96米（13英尺），仅尾翼就有8层楼那么高。这个家伙重达18.14吨（40万磅），螺旋桨长度超过27英尺。它的8个重达5000磅的28缸引擎加起来有多达17000千瓦，能加6.36万升（1.4万加仑）燃油。更不同寻常的是，由于战时条件有限，它是由木头造的，这也是它绰号的由来。事实上，"云杉鹅号"主要是用桦木而不是用云杉造的。

由于要花费大量时间研究如何制造如此巨大的飞机，再加上供给问题和休斯的极端完美主义，因此制造工作进展缓慢。1944年，凯撒审时度势，退出了这个项目。直到战争结束，这架被正式命名为H-4的飞机还是未能如期制造出来。但是休斯毫不在乎，他又自掏腰包，拿出700万美元投入这架飞机的制造中，即使当时战争已经结束，不再需要这样的飞机了。1946年夏天，休斯在试飞为政府试制的XF-Ⅱ时，飞机坠毁于一座山庄，休斯身受重伤。这不是他唯一遇到的困难，政府开始质疑所有的钱到底用在了什么地方，为什么四年都未能交付一架飞机。

1947年，美国参议院组成了一个战时调查委员会，开始对休斯的公司进行听证，还涉及对好莱坞新人潜规则的指控。休斯出现在委员会面前，肆无忌惮

a history of sleazy① backhanders② involving Hollywood starlets. When Hughes appeared in front of the committee he brazened③ it out saying:

> The Hercules was a monumental undertaking. It is the largest aircraft ever built. It is over five stories tall with a wingspan longer than a football field. That's more than a city block. Now, I put the sweat of my life into this thing. I have my reputation all rolled up in it and I have stated several times that if it's a failure I'll probably leave this country and never come back. And I mean it!

But Hughes had been seriously stung by critics who described his beloved creation as a 'flying lumberyard④' and taunted⑤ him that it would never fly. He loathed⑥ its derogatory⑦ Spruce Goose nickname. Hughes decided to show them all. During the hearings he had the only Goose ever completed at last delivered to Long Beach, California, and readied the craft for its first flight.

On 2 November, with Hughes himself at the controls, the plane skimmed across the water before suddenly gliding⑧ into the air for the first time. Hughes flew it at just 70ft for a total of 1 mile. It wasn't much, but Hughes had proved that the Spruce Goose could, indeed, fly. The investigation into Hughes' affairs never really got anywhere, but the government still cancelled its order for the plane.

Even by 1948 it was clear that Hughes, an obsessive⑨, hadn't given up. A glowing report in *Popular Mechanics* reckoned⑩ the plane 'is slated⑪ to be for many years the biggest plane that ever flew'. Sadly it was also destined to be the biggest plane that never flew again. Still Hughes, who subsequently became a recluse⑫, could never let go. For the next quarter of a century

注释

① sleazy ['sliːzɪ] *adj.* 肮脏的，低级庸俗的

② backhander ['bækhændə] *n.* 贿赂，反手一击

③ brazen ['breɪz(ə)n] *vt.* 使变得无耻，厚脸皮地做

④ lumberyard ['lʌmbəjɑːd] *n.* 木材堆置场，储木场

⑤ taunt [tɔːnt] *vt.* 奚落，逗弄

⑥ loathe [ləʊð] *vt.* 讨厌，厌恶

⑦ derogatory [dɪ'rɒɡət(ə)rɪ] *adj.* 贬损的

⑧ gliding ['ɡlaɪdɪŋ] *adj.* 滑行的，流畅的，滑顺的

⑨ obsessive [əb'sesɪv] *adj.* 强迫性的，着迷的

⑩ reckon ['rek(ə)n] *v.* 测算，估计，认为

⑪ slate [sleɪt] *vt.* 严厉批评某人

⑫ recluse [rɪ'kluːs] *n.* 隐士，隐居者

地说：

> "大力神号"是一项不朽的事业。它
> 是史上最大的飞机，比五层楼还要高，翼展
> 比足球场还要长，那简直比一个街区还大。
> 现在我将毕生的心血都投入这架飞机中。我
> 的声誉全在于此，我也几次说过，如
> 果它失败了，我将离开美国，永不回
> 来。我是认真的！

但是评论家的话深深刺痛了休斯，他
们把他所钟爱的发明说成是"飞行的木料
场"，嘲讽说永远不可能飞起来。休斯讨厌
"云杉鹅号"这个略含贬义的绰号，他决定
飞给他们看看。听证会期间，他把这架好不
容易完成的唯一的"鹅"运往加利福尼亚州的长堤海
岸，并为第一次飞行做好了准备。

1947年11月2日，在休斯的亲自驾驶下，这架飞
机掠过水面，然后猛地滑向空中，实现了首飞。休斯
以仅21.34米（70英尺）的高度飞行了1.61千米（1英
里）。虽然没有多厉害，但是已经证明"云杉鹅号"
确实可以飞起来。关于休斯的调查虽没有取得任何进
展，但政府仍然取消了飞机的订购合同。

甚至到1948年，执着的休斯仍然没有放弃。《大
众机械》杂志在一篇报道中热情洋溢地评价这架飞
机"是飞行过的最大飞机，这个纪录将会保持很多

▲ 1947年，霍华德·休斯的"大
力神号"H-4的第一次（也是
仅有的一次）飞行。这架飞机
被称为"云杉鹅号"，是世界
上最大的飞机，而且至今仍然
保持着最长翼展纪录。

*Howard Hughes' H-4 Hercules on
its maiden (and only) flight in 1947.
Dubbed the 'Spruce Goose', it was
the world's largest plane and still
holds the record for the longest
wingspan.*

he kept it in a specially designed, climate controlled hanger at a cost of $1 million a year.

After his death in 1976 no one quite seemed to know what to do with it. Even Disney didn't want it. Finally, in the 1990s, the Goose found a buyer who wanted to make it part of his own collection of wartime aircraft. And, in 2001, the fully restored dinosaur went on display at its new home, the Evergreen Aviation Museum in Oregon. There the plane remains as a fitting memorial to a man who just didn't know when to stop.

年"。不幸的是,它也注定是一架只飞行过一次的最大的飞机。休斯后来过起了隐居生活,但仍然没能放下这架飞机。后来的25年中,他每年花费100万美元把这架飞机保存在专门设计的有空调的吊架上。

1976年休斯去世后,似乎没人知道该怎么处理这架飞机,连迪斯尼乐园都不想要它。直到20世纪90年代,最终有人愿意买这只"鹅",作为战时飞机的收藏品之一。2001年,这架重新组装的"云杉鹅号"飞机开始在它的新家——俄勒冈州的长荣航空博物馆展出。在那里,这架飞机仍然是对一位永不放弃的人的最好纪念。

DISASTER
EXPLODING TRAFFIC LIGHTS

Today, London pavements are clogged① with tourists photographing Big Ben or waiting at the modern traffic lights to cross the road for a better look at the famous landmark. But most of those who linger② momentarily at the corner of Bridge Street and Whitehall are oblivious③ to the tiny blue plaque④ on the wall above and behind them. It recalls the almost forgotten historic importance of the spot where they stand. For this is the sight of the world's very first set of traffic lights, erected⑤ way back in 1868 – a date when the Houses of Parliament across the street were only just receiving their finishing touches.

Traffic lights have, of course, become a norm of modernday life and a crucial part of any town-planner's armoury⑥. Love them or loathe⑦ them, they save lives and help us manage our traffic. Yet they could have been commonplace⑧ fifty years before they began to appear routinely at busy junctions. Sadly, a chance tragedy would kill off John Peake Knight's revolutionary idea before it was properly put to the test. Knight's gas-fired traffic light was to prove simply too ahead its time, falling foul⑨

爆炸的红绿灯

一场灾难

今天的伦敦街道，观光游客络绎不绝，他们要么在给大本钟拍照，要么在等红绿灯，以便穿过马路更好地一睹这座著名的伦敦标志性建筑。然而，在大桥街和白厅一角短暂驻足的多数游客并没有注意到身后墙上一块极小的蓝色铭牌，这块铭牌记录了曾经发生在这里的一段几乎被忘却的历史。这里是世界上第一盏红绿灯安装的地方，当时是1868年，街对面的英国议会大厦刚刚竣工不久。

毋庸置疑，红绿灯已成为现代生活中不可或缺的一部分，已成为城市规划者需要考虑的重要部分。无论我们喜欢也好，厌恶也罢，红绿灯都挽救了无数生命，使交通井然有序。然而就在红绿灯开始普遍出现于车水马龙的路口的50年之前，它们本来可以普及。遗憾的是，这个还未经实践检验的革命性想法，就被一次

▲ 大桥街和白厅一角的蓝色铭牌。

The blue plaque at the corner of Bridge Street and Whitehall.

注释

① lobby ['lɒbɪ] *n.* 大厅，休息室，会客室
② demise [dɪ'maɪz] *n.* 死亡，传位
③ gadget ['gædʒɪt] *n.* 小玩意儿，小器具
④ superintendent [sju:p(ə)-rɪn'tend(ə)nt] *n.* 监督人，负责人，主管
⑤ obituary [ə(ʊ)'bɪtʃʊərɪ] *n.* 讣告
⑥ Derby Day 德比赛马日
⑦ Select Committee （议会中为调查某问题而设立的）特别委员会
⑧ metropolis [mɪ'trɒp(ə)lɪs] *n.* 大都会，首都

of the technology then available to him and pressure from the Victorian health and safety lobby①. Incredibly, the unlucky demise② of his gadget③ meant that no one else would try and introduce a traffic light to any street corner anywhere in the world until well into the twentieth century.

Even before the age of the petrol engine, however, some sort of traffic flow device was badly needed. In Victorian London the roads were clogged with horse-drawn carriages and arguably almost as dangerous as they are in today's age of the motor car. During 1866 a total of 102 people were killed on the capital's roads. In 2009 the figure was 184.

J.P. Knight, a Nottinghamshire railway engineer, believed he had the answer to the growing death toll on the roads. Born in 1828, he left school at just 12 and worked his way up to become traffic superintendent④ on the South Eastern Railway. His obituary⑤ later told of how he was particularly good at 'managing the crowds on Derby Day'⑥ and was instrumental in the adoption of new braking systems for trains, as well as emergency bell-pulls for lone female travellers in train carriages. Concerned too about the state of traffic management on the roads he wrote to the Home Secretary recommending one way streets. His letter read: 'Narrow streets, where two vehicles cannot well pass each other should be used for traffic in one direction only, so as to prevent vehicles meeting.'

Then, in December 1868, two years after his idea for traffic signals in central London went before a Parliamentary Select Committee⑦, the authorities gave the go ahead. *The Express* reported at the time:

The regulation of the street traffic of the metropolis⑧,

偶然的悲剧葬送了。它的发明者是约翰·皮克·耐特（John Peake Knight）。事实上，耐特的煤气红绿灯只是太超前，当时的技术无法实现他的想法，而且还得承受着维多利亚时期健康安全游说的压力。他的发明被束之高阁，直到20世纪才有人再次尝试把红绿灯引入全世界的各条大街上，这真让人难以置信。

尽管如此，在汽油发动机出现之前，人们就迫切需要一种控制交通流的装置。维多利亚时期的伦敦街道常常因为马车所堵塞，几乎和现在的汽车一样危险。1866年就有102人因交通事故在伦敦大街上丧生，而2009年则有184人丧生。

耐特（J.P.Knight）是诺丁汉郡的一名铁路工程师，他认为自己找到了解决街道死亡人数不断攀升的方法。耐特生于1828年，12岁辍学，后来成为东南铁路公司的交警。通过他的讣告，我们知道他特别擅长"管理德比赛马日的拥挤人群"，还曾协助发明了一种新式火车刹车系统，还为火车车厢内的单身女性乘客设计了紧急拉铃装置。出于对道路交通管理状况的关心，他写信给内政部部长建议施行单行道。他在信中写道："对于两辆车不便同时通过的狭窄街道，应当只允许单向通行，以避免会车。"

1868年12月，也就是耐特关于在伦敦市中心使用交通信号灯的建议被呈递至议会专责委员会讨论两年后，当局批准了这个建议。当时的《每日快报》是这样报道的：

注释

① semaphore ['seməfɔ:] n. 信号标，旗语

② hitherto [hɪðə'tu:; 'hɪðətu:] adv. 迄今，直到某时

③ gesticulation [dʒəˌstɪkjʊ'leɪʃ(ə)-n] n. 手势，姿势，示意动作

④ proceed [prə'si:d] v. 接着做，继而做

⑤ illuminate [ɪ'l(j)u:mɪneɪt] vt. 阐明，说明，照亮

⑥ lever ['li:və] n. 杠杆，控制杆

⑦ keen [ki:n] adj. 敏锐的，渴望的
be keen on 喜爱，渴望

⑧ fro [frəʊ] adv. 向后，向那边

⑨ archive ['ɑrkaɪv] n. 档案馆，档案文件

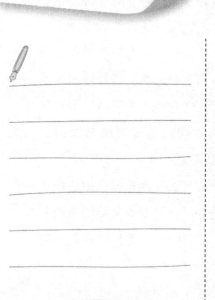

the difficulties of which have been so often commented upon, seems likely now to receive an important auxiliary. In the middle of the road between Bridge Street and Great George Street, Westminster, Messrs Saxby and Farmer, the well known railway signalling engineers, have erected a column 20 feet high, with a spacious gas lamp near the top, the design of which is the application of a semaphore① signal to the public streets at points where foot passengers have hitherto② depended for their protection on the arm and gesticulation③ of a policeman – often a very inadequate defence against accident.

As this indicated, the design of the Knight's traffic lights, which relied heavily on train signals, featured two semaphore style arms. When they were raised it meant traffic should stop. When lowered that they could proceed④ with caution. At night-time red and green gas lights would also be illuminated⑤ thanks to power from a gas pipe running up the middle of the post. The whole thing was to be operated by a policeman with a lever⑥ with the arms raised to stop traffic for thirty seconds in every five minutes. Wags joked that the Home Office was keen on⑦ the idea because it would protect the lives of MPs and officials going to and fro⑧ Whitehall.

On 10 December the first set of traffic lights were installed at the junction where traffic poured off Westminster Bridge and met the throng coming down Whitehall. Made from cast iron and painted green and gold, the typically grand Victorian devices were even topped with a pineapple finial. A poster, now in the Metropolitan Police archive⑨, was put up to inform travellers about the new device. It read:

鉴于首都街道交通管理之难，饱受批评甚多，应当引入一种重要的辅助设备。著名铁路信号工程师萨克斯比和法默先生在威斯敏斯特区大桥街和大乔治街之间的道路中间立起了一根20英尺高的柱子，柱子上方有一个大容量煤气灯。这一设计把信号灯应用于公用街道，可以像警察的手臂和手势一样保护行人，但比警察的指挥更为妥当。

这篇文章表明，耐特红绿灯的设计主要以铁路信号为依据，具有两条信号灯臂。信号灯臂升起表明车辆应当停止，降下表明车辆可以谨慎通行。在夜间，煤气通过煤气管到达灯柱中间，点燃红灯和绿灯。这套装置由一名警察通过一根杠杆进行操作，每5分钟抬起信号灯臂使车辆停止30秒钟。爱开玩笑的人说内政部之所以热衷于这个想法，是因为它可以保护往来于白厅的议员和官员的性命。

1868年12月10日，威斯敏斯特大桥交通路口安装了第一盏红绿灯。在那里，威斯敏斯特桥上的车马奔流而下，与从白厅蜂拥而出的人群相遇。这套装置由铸铁制成，涂成金绿色，顶部甚至还有菠萝形饰物，颇具典型的维多利亚风格。为了让行人了解这套装置，路口贴出了一张布告。这张布告现在保存于伦敦警务处档案室，其内容如下：

By the signal 'caution', all persons in charge of vehicles and horses are warned to pass over the crossing with care and due regard to the safety of foot passengers. The signal 'stop' will only be displayed when it is necessary that vehicles and horses shall be actually stopped on each side of the crossing, to allow the passage of persons on foot; notice being thus given to all persons in charge of vehicles and horses to stop clear of the crossing.

Proclamation[1] of Richard Mayne, London Police Commissioner

In fact, a staggering 10,000 of these leaflets[2] and posters were put up all over London.

The Express enthused[3]: 'A more difficult crossing place could scarcely be mentioned and should the anticipations[4] of the inventor be realised similar structures will, no doubt be speedily erected in many other parts of the metropolis'. Indeed, plans were soon afoot[5] to install the signals on other busy roads in the capital, like Fleet Street. Then, without warning, on 2 January, the traffic light exploded. The policeman operating it was badly burned. Some reports suggest he later died. The injuries were certainly bad enough for the traffic light to be quickly taken down and the idea was completely dropped. The gas which fuelled the lights had proved too dangerous.

Ironically, Knight also worked on the first electrically lit train carriages. And only with the advent[6] of electricity was a similar idea to his traffic lights eventually revived in 1912 by the aptly[7] named Lester Wire, a police officer in America's Salt

注释

① proclamation [prɒklə'meɪʃn] *n.* 公告，宣告
② leaflet ['liːflət] *n.* 传单
③ enthuse [ɪn'θjuːz; en-] *v.* 充满热情地说，热烈地讲
④ anticipation [æntɪsɪ'peɪʃ(ə)n] *n.* 预料，预期
⑤ afoot [ə'fʊt] *adj.* 在进行中的，准备中
⑥ advent ['ædvɛnt] *n.* 到来，出现
⑦ aptly ['æptli] *adv.* 适宜地，适当地

当信号为"小心"时，所有过往车辆马匹须小心行使，注意行人安全；当信号为"停止"时，路口两端的车辆马匹必须停止，保障行人安全通行；出现该信号时，所有车辆和马匹应在路口外停止。

伦敦警察局局长理查德·梅恩宣布

事实上，这样的传单和布告在全伦敦共张贴了1万多份，真是惊人！

《每日快报》热情洋溢地评论道："这儿是伦敦最复杂的路口了，如果发明者的愿望能够实现，那么肯定会很快在首都的其他地方安装类似设备。"的确，安装红绿灯的计划在伦敦的其他繁忙路段也开始实施了，譬如舰队街。接着，1869年1月2日，在毫无征兆的情况下，红绿灯爆炸了。操作红绿灯的警察被烧成重伤，有报道说他后来因伤致死。事故发生后，红绿灯很快就被拆掉了，这个想法也彻底破产。事实证明，用煤气作为灯的能源太危险了。

具有讽刺意味的是，耐特也曾经研究过第一款用电照明的火车车厢。直到1912年，随着电气时代的到来，美国盐湖城一位名叫莱斯特·怀尔（Lester Wire）的警官逐渐用电实现了耐特的红绿灯的想法。伦敦则到1926年才再次使用红绿灯，这次是安装在圣詹姆斯街和皮卡迪利街的交叉口。这时红绿灯仍然需

▲ 1868年警方布告提醒伦敦人新煤气红绿灯的意义。这个装置在伦敦只存在了很短的时间，不久发生的悲剧就结束了这个试验。

Police poster from 1868 warning Londoners of the new gas-powered traffic lights which briefly appeared in the capital, before a tragedy put an end to the experiment.

注释

① Piccadilly [ˌpikəˈdili] *n.* 皮卡
迪利大街（伦敦的繁华街道）
② wreath [riːθ] *n.* 花冠，圈状物

Lake City. It wasn't until 1926 that London was again to get its next set of traffic lights – put up at the junction of St James's Street and Piccadilly①. Even this still had to be operated by a policeman before automated lights finally came in on 14 March 1932.

Knight himself died in 1886 following a stroke. Some 2,000 people turned out for his funeral procession and a wreath② was sent by the Prince of Wales. He was just 58. Who knows? Had the genius of his invention been properly recognised, tested and perfected the premature deaths of many more might have been saved.

要一名警察手动操作，直至1932年3月14日，自动红绿灯才最终投入使用。

耐特本人于1886年死于中风，年仅58岁。约有2000人参加了他的葬礼，威尔士亲王还献上了一个花圈。假如他的天才发明得到应有的认可、试验和完善，也许就不会有那么多的人过早地因交通事故而离开人世。

FAILED
THE STEAM-POWERED PASSENGER CARRIAGE

For an enterprising① soldier eager to advance, little beats inventing an improved way of killing people. So when Nicholas Cugnot designed a new rifle at the very time the French General of Artillery②, Jean Baptiste Vaquette de Gribeauval, was working on his own system for standardising weapons, he endeared③ himself into the higher echelons④ of the military. Cugnot's rifle was soon part of the French army's arsenal⑤ and, impressed with the young engineer, Gribeauval found further work for him. It was to lead to the very first powered car, the very first car accident and, quite possibly, the very first arrest for dangerous driving. The steam-powered car was on the way, and record breaking though it was, it wouldn't stand the test of time.

Having already revolutionised French weaponry, in the 1760s General Gribeauval set to work improving the way munitions⑥ were moved. Horses, the traditional form of transport in the eighteenth century, had been hauling⑦ cannons and guns since the dawn of warfare. But, at a time when British industrialist James Watt was improving the steam engine, this was the new age of transport. With an engineer in his ranks

蒸汽客车

对于具有上进心、事业心的士兵来说，再没有什么比发明一种更好的武器更具有吸引力的了。所以当尼古拉斯·居纽（Nicholas Cugnot）设计出一种新型步枪时，法国炮兵将军让·巴普蒂斯特·瓦克特·德·格利博瓦尔（Jean Baptiste Vaquette de Gribeauval）正在研究自己的武器标准化体系，居纽由此受到军队高层的器重，不久居纽设计的步枪就被收入法国军队军械库。格利博瓦尔觉得这位年轻工程师颇有才华，就赋予他一项新任务，结果产生了世界上第一辆动力车、第一起动力车交通事故，甚至极有可能是第一起因危险驾驶而遭到拘捕的案例。蒸汽动力车就这样出现了，尽管它的发明具有突破性，却未能经得住时间的考验。

18世纪60年代，格利博瓦尔将军改革了法国武器体系之后，开始着手改进军火运输方式。早在人类战争之初，马就被用来运送大炮和枪械，直到18世纪，马仍然是传统的运输工具。然而，英国工业家詹姆斯·瓦特（James

注释

① combustion [kəm'bʌstʃ(ə)n] *n.* 燃烧

② rotational [ro'teʃənl] *adj.* 转动的，回转的，轮流的

③ piston ['pɪst(ə)n] *n.* 活塞

④ counterweight ['kaʊntə,weɪt] *n.* 平衡力，平衡物

⑤ armament ['ɑːməm(ə)nt] *n.* 武器，军备

⑥ cumbersome ['kʌmbəs(ə)m] *adj.* 笨重的，累赘的，难处理的

⑦ tractor ['træktə] *n.* 拖拉机，牵引机

⑧ cabin ['kæbɪn] *n.* 小屋，客舱

⑨ synchronise ['sɪŋkrənaɪz] *v.* 同步，同时发生

⑩ reciprocating [rɪ'sɪprə,keɪtɪŋ] *adj.* 往复的，交互的

⑪ axle ['æks(ə)l] *n.* 车轴

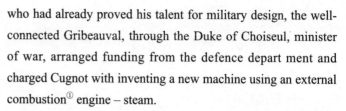

who had already proved his talent for military design, the well-connected Gribeauval, through the Duke of Choiseul, minister of war, arranged funding from the defence depart ment and charged Cugnot with inventing a new machine using an external combustion① engine – steam.

By 1769 Cugnot had patented the world's first 'motorised carriage', basically a boiler on three wheels that could reach speeds of 2.5 mph by converting the lateral energy of steam into rotational② energy to power wheels. His breakthrough was placing the engine over the front wheel, and then using two pistons③ attached to each wheel to propel the vehicle. With a driver on board to steer the unstable, weighty vehicle in as much of a straight line as possible (changing direction was a major operation), as well as stoke up the boiler, the car was born. But there were problems. With the boiler at the front, it had a tendency to tip onto its nose unless counterweighted④ by a canon at the back – although as its mission was to tow 5 tons of armament⑤ this wasn't necessarily a bad thing. Of more concern, it could only produce enough steam to run for fifteen minutes at a time. And there wasn't a great deal of space to carry spare fuel or water. In short, this was a vehicle that most people could out-walk. But it was a start.

Cugnot tried again. The first ever mechanically propelled vehicle may have been cumbersome⑥, more tractor⑦ than car, but it worked. What if he could improve speed and reach and make the cabin⑧ larger so that it could take passengers? Within a year, with his colleague, army mechanic Michel Brezin, and with the support once more of the duke and the general, Cugnot's next steam vehicle accommodated four people. As with the earlier version, two pistons connected by a rocking beam were synchronised⑨, so when atmospheric pressure forced one down, the other went up, creating reciprocating⑩ motion through the axle⑪ to turn the wheels. But this time, the pistons

Watt）对蒸汽机的改进标志着运输新时代的到来。手下有这么一个军事设计的天才，人脉很广的格利博瓦尔通过战争部长舒瓦瑟尔（Choiseul）公爵从国防部获得经费，任命居纽用外燃机（即蒸汽机）发明一种新机器。

　　1769年，居纽已经申请了世界上第一辆"机动马车"的专利，其基本原理是一辆带锅炉的三轮车，通过将蒸汽的横向能量转化为转动能量，然后为轮子提供动力而行驶，时速可达4千米（2.5英里）。他的突破性成就在于将发动机置于前轮，通过装在每个车轮上的两个活塞驱动马车前进。司机在车上控制这辆不稳的、笨重的车子，使其尽量保持直线行驶（改变方向是主要操作），还要给锅炉添加燃料。汽车就这样诞生了，不过还是存在不少问题。由于锅炉位于车身前部，很容易前倾，除非在车身尾部用一门大炮来平衡，因为车子本来就是用于牵引5吨重的武器装备的，所以这还算不上一件坏事。更严重的问题是蒸汽机一次产生的蒸汽只够车子行驶15分钟，而且没有足够空间装载备用燃料或水。总之，这辆车走起路来还不如一般人。不过这毕竟只是一个开始。

　　居纽再次尝试。这辆史上第一辆机动车或许有点儿笨重，看上去更像拖拉机而不像汽车，但它能够正

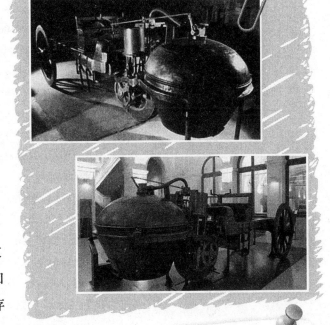

▲ 尼古拉斯·居纽设计的"机动马车"在巴黎现代艺术博物馆展出。他的"轮子上的锅炉"于1769年完成首次行驶。

Nicholas Cugnot's fardier à vapeur on display at the Musée des Arts et Métiers in Paris. His 'boiler on wheels' made its first journey in 1769.

注释

① condensation [kɒnden'seɪʃ(ə)n]
n. 冷凝，凝结，压缩
② furnace ['fɜːnɪs] *n.* 火炉，熔炉
③ trundle ['trʌnd(ə)l] *v.* （使缓慢、
轰鸣地）移动，滚动
④ perpetuator [pər'petʃueɪtər] *n.*
永久保存的人
⑤ hedge [hedʒ] *n.* 树篱，障碍
⑥ ditch [dɪtʃ] *n.* 沟渠，壕沟
⑦ slur [slɜː] *n.* 污点，诽谤
⑧ collision [kə'lɪʒn] *n.* 碰撞，冲
突，（意见，看法）的抵触
⑨ pension ['penʃ(ə)n] *v.* 发给养老
金或抚恤金

could be moved without condensation① from the high-pressure steam engine, thus increasing efficiency. And the vehicle was in two parts – at the front, a copper boiler with a furnace② inside and two small chimneys above, and at the back the carriage on two wheels for the passengers and a seat at the front for the driver.

The duke and the general, together with other French military dignitaries watching the trial, were impressed. No longer did the driver have to stop every fifteen minutes to add water, then sit around waiting for it to boil. The vehicle, which could now travel for an hour and a quarter without a break, was still harder to operate than a horse-drawn carriage and, with the invention of brakes still years away, was all but impossible to stop. Nevertheless, approval was granted for further development.

Sadly, in his quest for greater efficiency, brakes weren't top of the list for Cugnot. One day in 1771, while trundling③ along at a speed of up to 3 mph in his mark two model *fardier* (French for dray, as the car was known), he became both victim, and perpetuator④, of the first car accident when, unable to brake, he steered the steamer into a wall. Some reports suggest that this was just the first in a series of accidents. The vehicle also overturned while attempting to turn a corner in the centre of Paris. Tired of picking him out of hedge⑤ and ditch⑥, the police eventually charged Cugnot with dangerous driving, although no court records have been found to prove this slur⑦ and the earliest recording of the accident is in Cugnot's 1804 obituary. Quite coincidentally, the end of Cugnot's motoring career came not long after this collision⑧. His military and political masters fell out of favour and Cugnot was pensioned⑨ off.

Self-propelled vehicles, however, continued to steam ahead. Within a few years, French engineer Onésiphore

常行驶。要是他能提高车速和行驶距离，扩大车厢使其能运载乘客，那将会怎样呢？一年之内，在舒瓦瑟尔公爵和格利博瓦尔将军的再度支持下，他和同事米歇尔·布莱仁（Michel Brezin）共同研制出了能乘坐四人的蒸汽汽车。这个型号与前一个型号一样，两个活塞以摆杆连接并实现同步，当气压迫使一个活塞下降时，另一个活塞就会上升，通过轮轴往复运动驱动车轮。不同之处在于，新蒸汽汽车的活塞运动时，高压蒸汽引擎不会发生冷凝作用，从而提高了效率。此外，车身被分为两部分：前半部分是一个铜锅炉，里面装有炉子，上面有两个小烟囱；后半部分是置于两个车轮上的乘客车厢，车厢前面是驾驶员座椅。

公爵和将军以及其他军队高官共同观看了试车之后，纷纷赞叹不已。这次司机无须隔15分钟就停车加水，等水烧开再继续前进。虽然汽车可以持续行驶1小时15分钟，但还是比马车难操作，加之没有刹车（那是多年以后的事情了），根本不可能立刻停车。然而无论如何，它还是获得了进一步研发的许可。

遗憾的是，居纽专注于追求更高效率，而忽视了对刹车的研究。1771年的一天，当他驾驶着"法第尔二号"（法第尔是法语词，意为运货马车）以4.8千米/时（3英里/时）的速度笨重地行驶时，由于无法刹车，蒸汽汽车撞上了墙。他由此成为第一起机动车交通事故的受害者和肇事者。有报道称，这只是他制造的一系列交通事故中的第一起事故。他的汽车在巴黎市中心转弯时还翻了车。警察厌倦了一次次地把居纽从篱笆和水沟中救出来，最后以危险驾驶的罪名把他告上法庭。不过人们并没有找到有损他

Pecqueur invented the differential gear, which allowed power to be transferred more efficiently to car wheels. In America, steamboats began to stay afloat, a dramatic improvement on the very first one in 1763, when William Henry's aqua^① invention sank. And by the early 1820s, steam-powered stagecoaches^② taught Britons how small the country really was. In the fine spirit of British progress, they were quickly outlawed because of the noise and danger.

If Nicholas Cugnot was the first person to get passengers moving, Father Ferdinand Verbiest, a Flemish mathematician and Jesuit missionary who moved to China to spread the good word, actually designed the first self-propelled steam car – but was unable to progress further than building a scale model. His place in history is therefore as the creator of the very first powered toy car – more than a hundred years before Cugnot invented it for real. Somewhere between 1671 and 1679, Verbiest used steam to power a small turbine^③ to turn the wheels of a vehicle. This work of genius he gave to the Chinese emperor, whom he had much to thank for. Jesuit missionaries in China didn't always go down terribly well and, a few years earlier, after being found guilty of teaching a false religion, Verbiest had been sentenced to death by a thousand cuts, China's most tortuous^④ method of execution. Fate intervened when an earthquake destroyed the execution site and, as the authorities took this to be an omen^⑤, Verbiest instead ended up working for the emperor on redesigning the Chinese calendar (he ordered them to take out a month) and designing toy cars. When the steam vehicle powered across the floor, turned around and shuddered^⑥ back to its owner, the court was delighted. The omen had been a good one.

However, technology inevitably put paid to the external combustion engine in cars. Steam vehicles, being heavy, were,

注释

① aqua ['ækwə] *n.* 水, 溶液, 浅绿色
② stagecoach ['steɪdʒkəʊtʃ] *n.* 公共马车, 驿站马车
③ turbine ['tɜːbaɪn] *n.* 涡轮机, 汽轮机
④ tortuous ['tɔːtʃuəs] *adj.* 拐弯抹角的, 含混不清的, 冗长费解的
⑤ omen ['əʊmən] *n.* 预兆, 征兆
⑥ shudder ['ʃʌdə] *v.* 发抖, 战栗

声誉的法庭记录，而有关居纽最早的交通事故则记录在他1804年的讣告中。碰巧的是，这次撞车事故之后不久，居纽发明机动车的事业也走到了尽头。由于他所效力的军方和政治主人受到冷落，居纽也被迫退休。

尽管如此，机动车的发展仍旧蒸蒸日上。几年后，法国工程师奥内西佛尔·培克尔（Onésiphore Pecqueur）发明了差速齿轮，这种齿轮可以使动能更加有效地传递到车轮。威廉·亨利（William Henry）在1763年发明了第一艘汽船，但沉掉了，美国人在此基础上做了很大改进，汽船也由此发展起来。到了19世纪20年代早期，蒸汽动力公共汽车让英国人认识到了国土的狭小。然而，英国人具有追求进步的优良作风，由于噪声大、风险高，蒸汽动力公共汽车很快就被宣布为不合法。

如果尼古拉斯·居纽是让蒸汽客车动起来的第一人，那么到中国传播天主教的耶稣会士、佛兰德数学家南怀仁（Ferdinand Verbiest）神父则是设计蒸汽动力汽车的第一人，不过他的发明仅仅是一个与实物成比例的模型。因此，他发明了历史上第一辆机动玩具车，比居纽发明真正的蒸汽汽车早了100多年。1671—1679年，南怀仁曾使用蒸汽作为动力推动小型涡轮运动，从而使车轮转动起来。他把这件天才之作贡献给当时的中国皇帝，以表达他的感激之情。事实上，传教士在中国的传教之路并非一帆风顺：早些年，朝廷因南怀仁传播邪教而判他凌迟处死，这是中国古代最残忍的一种死刑。也许是命不该绝，一场地震摧毁了刑场，官府认为这是天意。于是，南怀仁被免去死罪，南怀仁还奉诏为皇帝重新修订中国历法（他让人

注释

① condenser [kən'densə] *n.* 冷凝器，[电] 电容器
② fossil ['fɒsl] *n.* 化石
③ vehicular [vi:'hɪkjʊlə] *adj.* 供车辆等使用的，车辆的，运输工具的
④ misconception [mɪskən'sepʃ(ə)n] *n.* 错误认识，误解
⑤ de facto [ˌdeɪ 'fæktəʊ] *adj.* 实际上存在的（不一定合法）

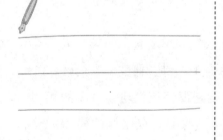

as Cugnot discovered, often slow. Constantly getting water to the boiler was trouble. Extra water, lots of it, has to be carried or a condenser① fitted, thus adding more weight. The engines were also noisy. When, in 1889, Karl Benz entered the fray with an internal combustion engine powered by fossil② fuel, for a while it looked like the battle for vehicular③ supremacy could go either way. But the greatest steam car of all, the Stanley Steamer, still had something to prove. Destroying the misconception④ that steam cars are always slow, in 1906 driver Fred Marriott, at the wheel of his Stanley Steamer, reached 127.659 mph (205.5 km/h); beating four petrol-powered cars to win the Dewer Trophy speed event and grab what is still the land speed record for a steam powered vehicle. It was the last major triumph for steam. Although petrol and steam cars were neck and neck in the race to become de facto⑤ standard, by the 1920s, the quieter, less troublesome internal combustion engine had won the battle and the steam ran out of the external combustion adventure.

们把中国历法减去一个月）和设计玩具汽车。当蒸汽汽车在地上前行、转弯，最后颠簸着返回主人身边时，皇宫里的人都乐了。这是个好兆头！

然而，随着科学技术的发展，外燃机逐渐退出了汽车的舞台。正如居组所见，蒸汽汽车太笨重了，所以速度往往很慢；问题是还得不停往锅炉里加水，这就需要携带大量水或者安装一台冷凝器，增加了汽车的重量；还有发动机噪声也非常大。1889年，卡尔·本茨（Karl Benz）把化石燃料动力内燃机引入汽车，使内燃机和外燃机之争一时变得难解难分。不过，最伟大的蒸汽汽车——斯坦利蒸汽汽车仍然要证明自身存在的价值。1906年，弗雷德·马里奥特（Fred Marriott）驾驶着斯坦利蒸汽汽车，以205.5千米/时（127.659英里/时）的速度，击败了四辆汽油动力车，创下了蒸汽汽车地面最高时速纪录，赢得了杜瓦奖杯速度赛，打破了蒸汽汽车速度慢的成见。这是蒸汽汽车的最后胜利。虽然汽油汽车和蒸汽汽车在成为行业标准的竞争中不相上下，但到20世纪20年代，噪声小、故障少的内燃机略胜一筹，外燃机退出了历史舞台。

▲ *路易斯·罗斯正在驾驶斯坦利蒸汽汽车（1903年）。*

Louis Ross racing a Stanley Steamer car, Florida, 1903.

FAILED
FLYING CARS

注释

① herald ['her(ə)ld] *v.* 是（某事）的前兆，预示
② feasible ['fiːzɪb(ə)l] *adj.* 可行的
③ cylinder ['sɪlɪndə] *n.* 圆筒，汽缸

Airborne automobiles were once the future; so thought motoring icon Henry Ford. His mass production methods, which led to the famous Model T, heralded① the expansion of car ownership in America in the first decades of the twentieth century. So people paid attention when, in 1940, he said: 'Mark my word. A combination airplane and motor car is coming. You may smile. But it will come.' In 2010 the USA alone had 136 million car owners and 187,000 civil aircraft. But the idea of combining the two is still mere fantasy. There isn't a flying car in every other suburban driveway and the skies above the world's major cities are not filled with winged private vehicles. It hasn't stopped a string of idealistic enthusiasts working on machines that you could use both on the highway and in the air. None have yet cracked the mass market as Ford expected, but he was certainly right in one sense: flying cars are perfectly feasible②.

Three years before Ford made his prediction, one Waldo Dean Waterman was already leading the way. His three-wheeled Arrowbile was built around a six-cylinder③ Studebaker car engine. It had no tail and the engine was placed at the rear of

飞行汽车

人们曾经以为将来总有一天会有飞行汽车，汽车大王亨利·福特（Henry Ford）也这么认为。他的大规模汽车生产技术，产生了大名鼎鼎的福特T型车，预示着20世纪前几十年美国的汽车持有率不断升高。于是人们记住了他在1940年说过的话："记住我的话，飞机和汽车就要结合起来了。也许你们会笑，但这一天一定会到来的。"到2010年，仅美国就有1.36亿人拥有汽车，民用飞机有187000架，但把两者结合起来仍然只是一个梦想。在郊区马路上看不到飞行的汽车，在世界各大城市的天空中也看不到飞翔的私家车。然而，许多热衷于飞行汽车的理想主义者研制空地两用汽车的脚步却始终没有停止。福特的预言还没有实现，还没有一辆飞行汽车进入大众市场，但他有一点绝对是正确的：飞行汽车在技术方面完全可行。

福特作出这个预言的三年前，沃尔多·迪恩·沃特曼（Waldo Dean Waterman）就已经开始研制飞行汽车了。他发明的三轮飞行汽车Arrowbile使用了一个6缸的斯图贝

the fuselage① along with the craft's propeller. To turn it from a car into a plane you essentially fixed the wings (with a span of 38ft) to the top when you wanted to fly. 'Add wings and it smoothly takes to the airways', ran the advertising copy. In August 1937 *Life Magazine* reported that the Arrowbile, later called the Aerobile, could fly at a speed of 120 mph, carry two passengers a distance of 350 miles and was yours for $3,000. On the roads it could do a sprightly② 56 mph. It could certainly fly and the Studebaker company ordered five. When it came down to it though, Waterman's figures didn't add up. His carcumplane turned out to be more expensive to make than it was commercially viable to sell and he didn't have the investment needed to see the plan through.

In 1946 the baton③ was taken up by Robert Edison Fulton Jr. who, during the war, had invented a simulator④-style machine for training aerial gunners. Flying himself across the US in a light plane for his work, he was frustrated at being stranded at airfields miles from the local town, unable to find a taxi or lift. He wondered why he couldn't just take his plane downtown. Despite having little expertise on aircraft design itself, Fulton nevertheless managed to build what he called the Airphibian. It was a simple but effective design, featuring a set of fabric⑤ wings and a tail that could be attached to the body of the car. Once removed you could drive from the airport back to your home and park the car in your garage⑥. The transformation from plane to car, or vice versa⑦, could all be done in five minutes flat. The Airphibian had a car's accelerator⑧ and brake⑨ pedal which doubled up as the plane's rudders in the sky. It could fly a distance of 400 miles, reaching speeds of 110 mph in the air and 55 mph on the ground. In 1950 the four-wheeled Fulton Airphibian became the first flying car to be certified for production by the US government's flying authority, the

① fuselage ['fjuːzəlɑːʒ] *n.*（飞机的）机身
② sprightly ['spraɪtlɪ] *adj.* 精力充沛的，精神矍铄的
③ baton [bə'tɑːn] *n.*（接力赛的）接力棒
④ simulator ['sɪmjʊleɪtə] *n.* 模拟装置
⑤ fabric ['fæbrɪk] *n.* 织物，布料
⑥ garage ['gærɑːʒ] *n.* 停车房，车库，汽车修理厂
⑦ vice versa [ˌvaisi'vəːsə] *adv.* 反过来也一样，反之亦然
⑧ accelerator [ək'seləreɪtə] *n.*（汽车等的）加速装置，油门
⑨ brake [breɪk] *n.* 刹车，制动器，车闸

克汽车引擎，没有尾翼，发动机与螺旋桨一同置于机身后部。当你想飞行时，只需要把机翼[翼展为11.58米（38英尺）]安装到车顶上，就可以使汽车变为飞机了。"装上翅膀，开始平稳飞行"，广告词如是说。1937年8月的《生活杂志》报道称，Arrowbile（后来又被称为Aerobile）能以193千米/时（120英里/时）的速度飞行，一次搭乘两名乘客连续飞行563千米（350英里）；在公路上能以90千米/时（56英里/时）的速度轻快地行驶，而它的售价仅为3000美元。沃特曼的汽车的确会飞，斯图贝克公司向他订购了5辆Arrowbile，但当真正准备生产这种飞行汽车的时候，沃特曼的数字没有继续增加。如果用于商业出售，这种飞行汽车的制造成本太昂贵了，他也没有足够的投资来完成这个项目。

1946年，发明飞行汽车的重任又落到了小罗伯特·爱迪生·富尔顿（Robert Edison Fulton Jr.）身上。富尔顿曾在战争期间发明了一种用于训练空中炮手的模拟设备。一次，他因公驾驶一架轻型飞机横穿美国，着陆后找不到出租车或顺风车，被困在距当地小镇几英里的飞机场，感到万分沮丧。他寻思为什么就不能直接把飞机开到城区呢？虽然富尔顿毫无飞机设计的专业知识，但他制造了名为Airphibian的飞行汽车。他的设计简单而高效，车身上安装了一套织布机翼和尾翼。把机翼和尾翼拆下来，即可驾车从机场回家，然后把车停在车库里。把飞机改装成汽车或把汽车改装成飞机只需要5分钟。Airphibian的汽车油门和刹车踏板可以叠合在一起，在飞行时可以充当飞机的方向舵。它可以连续飞行644千米（400英里），速

注释

① rigorous ['rɪg(ə)rəs] *adj.* 谨慎的，细致的，彻底的

② rescue ['reskjuː] *n.* 营救，援救

③ tow [təʊ] *v.* （用绳索）拖，拉，牵引，拽

④ scale [skeɪl] *n.* 规模；比例

⑤ rustle ['rʌs(ə)l] *vi.* （使）发出轻轻的摩擦声，发出沙沙声

⑥ dismayed [dɪs'med] *adj.* 担心的，失望的，忧虑的

⑦ prophesy ['prɒfɪsaɪ] *v.* 预告，预言

Civil Aeronautics Administration. The organisation itself even ordered ten of them at a cost of $7,500. Again money was the problem. The machine had been required to undergo rigorous① tests which meant Fulton was fast running out of cash and needed to bring in an investor. Disagreements with his backer meant Fulton eventually sold his own investment, abandoning the Airphibian altogether. Instead he went on to work on a much more successful device called the Skyhook, an aerial rescue② system for retrieving spies and soldiers from behind enemy lines, used by the US military for some thirty years.

Another flying car builder, also American, was a former missile tester called Moulton Taylor. He'd seen Fulton's car in 1946 and reckoned he could make a more practical version. His Aerocar, which did similar speeds to the Airphibian, had folding wings which were designed to be towed③ behind the car when it was on the road. After raising $50,000 to build and test a prototype in 1949, it took Taylor another seven years before he got his government certification. Finally he was ready to sell them – at $15,000 a time. In 1958 he said: 'My dream is to look up and see the sky black with Aerocars – and I'm sure that will happen someday.' A company called Ling-Temco-Vought seemed to be on board too, promising to go into full scale④ production if Taylor could rustle⑤ up 500 orders. When he only got half that number the company pulled the plug.

Taylor, who had given up a job in a military lab, was dismayed⑥ but not beaten. He went on refining his design throughout the 1960s, but without a factory in which to build his cars he struggled to produce cheap enough vehicles to meet the demand he felt existed. Then, in 1970, there was an ironic twist to the tale. Ford, the company whose founder had prophesied⑦ the rise of the flying car thirty years before, got in touch. The company's researchers reckoned they could sell 25,000 models

度可达177千米/时（110英里/时），地面行驶速度可达89千米/时（55英里/时）。1950年，富尔顿的四轮Airphibian成为获得美国政府飞行权威机构美国民用航空局许可生产的第一种飞行汽车。美国民航局甚至以7500美元的价格向他订购了10辆Airphibian。这次问题又出在钱上。批量生产前要求对Airphibian进行严格的测试，结果富尔顿很快就花光了现金，亟须找到一位投资者。然而，富尔顿与赞助者的分歧使他最终卖掉全部投资，放弃了Airphibian，转而去设计一种称为Skyhook的装置。这是一种用于从敌后营救间谍和士兵的空中救援系统，相对来说比较成功，美国军方使用了大约30年。

另一位飞行汽车设计者莫尔顿·泰勒（Moulton Taylor）也是位美国人，最初是一位导弹测试人员。他在1946年看过富尔顿发明的飞行汽车，自认为能设计出一种更实用的车型。他研制的Aerocar与Airphibian的速度相当，配有可折叠机翼，在公路上行驶时将机翼拖在车后。1949年，泰勒筹集了5万美元，制造并测试了一台样机，7年后他的发明得到了官方认证。最终他准备以每辆15000美元的价格出售这种汽车。1958年，他说："我的梦想是只要一抬头，就能看见空中到处都有Aerocar在飞行，我确信有朝一日这个梦想终将实现。"一家叫凌-特姆科-沃特（Ling-Temco-Vought）的公司似乎准备与泰勒合作，许诺如果泰勒能凑齐500个订单，他们就会大规模生产这种汽车。可是当他仅凑齐半数订单时，这家

of the Aerocar a year. For a moment the machine appeared to have a new lease① of life. But further research showed that simply to comply with tougher road and air regulation would cost some $400 million. Ford baulked② at the sum. In the end only six Aerocars were made.

Shortly afterwards, in the early 1970s, there was another bizarre attempt at a flying car which ended in spectacular③ failure. Engineer Henry Smolinski planned to go into production with his combined car and plane, the AVE Mizar, in 1974. On 11 September 1973 during a trial flight, the contraption came apart, killing Smolinski and the craft's pilot. Since then there have been many more flying cars. Many bold, clever and futuristic④ prototypes have emerged. But large-volume sales are still elusive⑤.

One of the problems with the first flying cars was that they were still pretty expensive for the man or woman in the street. Critics pointed to the fact that they also looked pretty ugly for something that was still a luxury item. Lighter engines and new materials now appear to make flying cars a better financial and technological bet⑥. But the main obstacle probably has never been the technology, so much as the general unease from big manufacturers and, above all, the authorities. The idea of managing millions of small aircraft whizzing⑦ about the

注释

① lease [li:s] *n.* 租约，租期
② baulk [bɔːk] *vi.* 犹豫
③ spectacular [spek'tækjʊlə] *adj.* 壮观的，惊人的
④ futuristic [ˌfjuːtʃə'rɪstɪk] *adj.* 未来派的，极其现代的
⑤ elusive [ɪ'l(j)uːsɪv] *adj.* 难懂的，易忘的，逃避的，难捉摸的
⑥ bet [bet] *n.* 打赌，赌注
⑦ whizzing ['wiziŋ] *v.* 发出嗖嗖声，嗖嗖掠过

公司就突然决定不生产这种汽车了。

这时泰勒已经辞去军队实验室的工作，他倍感失望，但并没有消沉。整个20世纪60年代他不断改进设计，但是没有一家工厂生产他的汽车。他竭力降低汽车成本，使汽车的价位能够为当时的市场所接受。1970年，故事发生了具有讽刺性的转折。福特公司联系了泰勒，30年前它的创始者曾预言飞行汽车的崛起。福特公司的研究者估计他们一年能售出25000辆Aerocar。一时间，这种汽车似乎又获得了新生，然而进一步研究表明，仅为遵守严格的公路和空中管制就得花费4亿美元左右。面对这么一大笔资金，福特公司改变了主意，最终只生产了6辆Aerocar。

不久之后，在70年代早期又出现一个怪异的飞行汽车发明，它的失败也让人唏嘘不已。工程师亨利·斯莫林斯基（Henry Smolinski）准备在1974年将他的汽车飞机结合体AVE Mizar投入生产。1973年9月11日，在一次试飞中，这个新奇的发明在空中解体，斯莫林斯基和飞行员在事故中身亡。这之后又有更多的飞行汽车被发明出来，涌现出许多大胆、精巧、充满未来主义色彩的设计，但最终仍未实现大规模销售。

第一代飞行汽车的一个问题是，对于普通消费者来说它们还是太贵了。而评论家认为，作为奢侈品，

▲ 莫尔顿·泰勒设计的Aerocar，首次开发于1949年，从未实现商业生产。

Moulton Taylor's Aerocar, first developed in 1949, never took off commercially.

注释

① conurbation [ˌkɒnɜːˈbeɪʃn] n. 有卫星城的大都市，（几个邻近的城市因扩建而形成一个）组合城市，集合都市

② minefield [ˈmaɪnfiːld] n. 布雷区，充满隐伏危险的事物

③ tailback [ˈteɪlbæk] n.（因堵车形成的）车辆长队

already crowded skies above the world's major conurbations① must seem daunting for any government. With that in mind, any company that looks to develop a flying car faces not only huge development costs but a bureaucratic health and safety minefield②. Flying cars might take pressure off our congested roads, but crashes could be pretty drastic affairs. And would they really free things up? Given that drivers would probably be forced to fly in narrow flight corridors, terrible tailbacks③ in the air would surely become commonplace too.

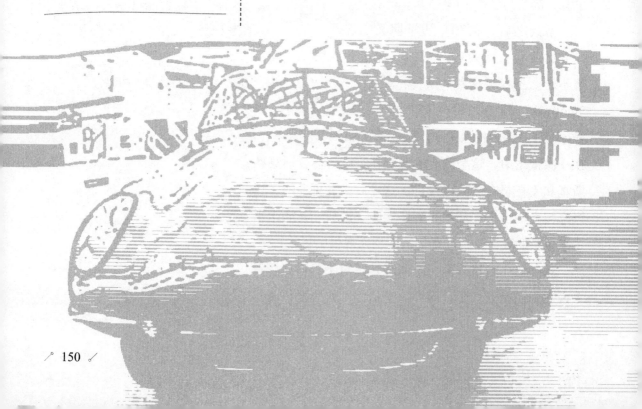

它们的外形又太难看了。如今出现了轻引擎和新材料，使制造飞行汽车在经济和技术上更加切实可靠，但普及飞行汽车的主要障碍也许并不是在技术上，而是来自大厂商和政府。试想要在那些本已拥挤的大都市上空管理数以万计快速飞行的小型飞行器，无论哪国政府都会感到头疼。因此，想开发飞行汽车的公司不仅要负担巨额的开发成本，还要小心应对来自官方的健康和安全问题。飞行汽车也许能缓解公路交通堵塞问题，但飞行汽车的坠毁问题也可能非常触目惊心。还有，它们是否真的能带来自由？如果驾驶员必须沿着狭窄的飞行通道行驶，空中的交通堵塞恐怕也会成为司空见惯的事。

POSTPONED
THE ATOMIC AUTOMOBILE

Imagine being able to clock up 5,000 miles in a state of the art vehicle without ever having to refuel①. That's the idyllic② vision of carefree motoring that the world's first nuclear powered car, the Ford Nucleon, promised to deliver. But it won't be coming to a showroom③ near you any time soon. And that might be just as well. Just consider the insurance payments on a car that could, potentially, lay waste to the very streets you're driving through.

Back in 1958, however, the idea of an atomic④ car promised to open up a brave new world of road transport. Scientists at Ford envisaged⑤ a time when mini nuclear reactors would be commonplace in everyday machines. In its glossy⑥ brochure to launch the idea of the Nucleon – surely the twentieth century's boldest car design – the company promised a 'glimpse into an atomic-powered future.' Its designers explained that such a car would, in time, be possible, because 'the present bulkiness⑦ and weight of nuclear reactors and attendant shielding⑧ will some day be reduced'.

While a full size Nucleon was never built, the firm did go

原子能汽车

试想你开着一辆最先进的汽车行驶8047千米（5000英里）而不用中途加油，那是何等场景？那种田园般的轻松驾驶是世界上第一辆核动力汽车——福特"核子号"（Ford Nucleon）所希望为我们带来的，但在你身边汽车展示厅里看到它的身影不是短时间内能实现的事。不过那样也好，如果你开的车可能会摧毁沿途的大街小巷，那你得支付多少保险费呀！

然而，追溯到1958年，原子能汽车本来有望为人们打开公路交通运输的一个美丽新世界。福特汽车的科学家设想，未来微型核反应堆将被广泛应用到日常机器之中。为了宣传福特"核子号"汽车——20世纪当之无愧的最大胆的汽车设计，福特公司印制了精美的小宣传册，让人们"一睹原子能提供动力的未来"。"核子号"的设计师解释说，这种汽车总有一天会成为现实，因为"这些庞大而笨重的核反应堆及其附带屏蔽装置将来会变得更为轻巧"。

注释

① tailfin [teɪlˈfɪn] *n.* 尾翼，垂直翼
② outlining [ˈaʊtˌlaɪn] *n.* 列提纲，描绘轮廓，提纲挈领
③ radioactive [ˌreɪdɪəʊˈæktɪv] *adj.* [核] 放射性的，有辐射的
④ converter [kənˈvɜːtə] *n.* 变流器，整流器，转化器
⑤ obsolete [ˈɒbsəliːt] *adj.* 淘汰的，废弃的，过时的
⑥ breezy [ˈbriːzɪ] *adj.* 通风良好的，有微风的，轻松愉快的
⑦ cantilever [ˈkæntɪliːvə] *n.* 悬臂，悬桁，伸臂
⑧ nasty [ˈnɑːstɪ] *adj.* 极差的，令人厌恶的，令人不悦的
⑨ meltdown [ˈmeltdaʊn] *n.* 核反应堆核心熔毁（导致核辐射泄漏）
⑩ prang [præn] *v.* 使（汽车）碰撞
⑪ awe [ɔː] *n.* 敬畏，惊叹
⑫ herald [ˈher(ə)ld] *v.* 是（某事）的前兆，预示

as far as constructing a 3/8 scale model, complete with classic 1950s tailfin① styling, as well as producing detailed publicity materials outlining② how it would work. Its designers intended to do away with the internal combustion engine. Instead the vehicle would be built with a power capsule containing a radioactive③ core element – sitting cosily in the car's boot. They explained that 'the drive train would be part of the power package and electronic torque converters④ might take the place of the drive train now used'.

Able to travel thousands of miles without refuelling, cars like the Nucleon would make fossil fuels obsolete⑤. When the Nucleon did finally need re-fuelling, you'd simply take it to a special recharging station. Owners could even order different sized reactors depending on how many miles you were likely to travel. The brochure continued breezily⑥: 'The passenger compartment of the Nucleon features a one-piece, pillar-less windshield and compound rear window and is topped by a cantilever⑦ roof.' There were other potential benefits, apart from not having to fill up. The car wouldn't need an ignition, it would probably be quieter to run than a traditional car engine and there would be no nasty⑧ emissions. Unless, of course, there was a crash! Not a word is included about the possible risks to the driver and passengers of a radioactive leak, or the potential havoc caused by a nuclear meltdown⑨ on the High Street after a prang⑩.

The concept of the Nucleon was born at a time when the popular belief was that nuclear energy could be utilised to solve many of society's problems and power almost anything. The era was dubbed the 'atomic age' and while many were awed⑪ by the destructive power of nuclear weapons, they also saw this incredible energy source as heralding⑫ a new industrial revolution. In 1954 Lewis Strauss, then chairman

虽然没有制造出一辆全尺寸的"核子号"，但福特公司制作了一个比例为3:8的"核子号"汽车模型，车尾为20世纪50年代经典尾翼造型，还印制了详细的宣传材料解释其工作原理。它的设计者不打算使用内燃机，而使用一个装有放射性核心物质的动力密封舱，它被舒适地安置在汽车后部的行李箱内。根据这些宣传资料，"驱动系统将成为动力装置的一部分，而电子变矩器有可能取代现有的驱动系统"。

"核子号"不用补充燃料也能行驶数千英里，这样的汽车让化石燃料失去了用武之地。当"核子号"需要补充动力时，你只要把它开到专门的反应堆更换站就可以了。车主甚至可以根据要旅行的远近选购不同大小的反应堆。宣传册上还这样轻快地写着："'核子号'的车厢由一块完整的无柱挡风玻璃、复合后窗玻璃和悬臂车顶组成。"当然，除了不用加油外，这种车还有其他优点。它无须点火，引擎也极有可能比传统汽车的引擎声音小，还没有令人讨厌的汽车尾气。当然，一种情况例外，那就是撞车！可是，对于司机和乘客可能面临的放射性泄漏危险，或者在大街上发生撞车事故后由于核爆炸而造成的巨大破坏，宣传册却只字未提。

▲ 福特"核子号"汽车的比例模型。这是一款20世纪50年代的概念车，使用一个微型核反应堆作为动力。

A scale model of the Ford Nucleon, a concept car from the 1950s, which was to be powered by a mini nuclear reactor.

of the United States Atomic Energy Commission, predicted that nuclear power offered the prospect of 'electrical energy too cheap to meter'. The first full-scale commercial nuclear power station opened at Calder Hall in Britain in 1956.

Within a few years, however, the public began to be more wary[1]. Concern grew about the limitations and dangers of nuclear devices, with a number of accidents at nuclear power plants. In America the Three Mile Island accident in 1979 was also an important milestone. The growing sense of unease may have been one reason why Ford didn't progress with its research into nuclear-powered cars. It's easy to see why. Aside from the safety aspects, the technological challenge and cost of developing the car would have been huge. There also seems to be no thought given to what to do with all the nuclear waste these cars would have produced.

While Ford were toying with atomic-powered cars, experiments were being made to use nuclear power in other methods of transport. Both the US and Soviet governments experimented with nuclear-powered aircraft in the 1950s. Such planes would have been able to stay in the air for very long periods of time. Improved ballistic[2] missiles eventually made the idea redundant; though even today, some scientists are debating the merits of nuclear-powered aircraft as one of the answers to the problem of global warming. Nuclear-powered[3] submarines[4], meanwhile, have been around since the USS *Nautilus*[5] put to sea in 1955.

While military applications of nuclear fuel are often considered safe enough to be used, it's not now likely that governments would allow

注释

① wary ['weərɪ] *adj.* 谨慎的，机警的
② ballistic [bə'lɪstɪk] *adj.* 弹道的，射击的
③ nuclear-powered ['njuːklɪə'pauəd] *adj.* 核动力的，核大国的
④ submarine ['sʌbməriːn] *n.* 潜艇
⑤ nautilus ['nɔːtɪləs] *n.* 鹦鹉螺，鹦鹉螺号

事实上，"核子号"构想产生于一定的时代背景。那时盛行这样一种观点，认为可以采用核能解决许多社会问题，可以给几乎任何东西提供动力。那个时代又被称为"原子时代"，虽然许多人对核武器的破坏性力量心有余悸，但他们相信这种不可思议的能源预示着一个新的工业革命的到来。1954年，美国原子能委员会主席刘易斯·施特劳斯（Lewis Strauss）预言，核能为我们呈现了一个"电能将便宜到无须计量"的未来。1956年，第一座大型商业核电站在英国考尔德豪尔开始运行。

然而几年后，公众对于核能的态度开始变得更加谨慎。核电厂发生的几次事故让人们进一步注意到核设备的局限性和危险性。1979年的三哩岛核事故更是一次具有里程碑意义的重大事件。福特公司在核能汽车研究上没有进展，也有可能是因为对核能的不安全感越来越强烈。原因显而易见：除了安全因素外，开发这种汽车还面临着巨大的技术问题和成本问题。此外，对于如何处理这种汽车所产生的核废料也似乎没有考虑。

当福特公司还在把玩原子能动力汽车时，其他人对别的交通工具也做了核能试验。20世纪50年代，美国和苏联政府都研制过核动力飞机，这种飞机能在空中飞行相当长的时间。不过，后来由于弹道导弹不断改进，人们不再认为研究核动力飞机有太大意义，尽管今天一些提倡使用核动力飞机的科学家认为这是解决全球气候变暖的一个办法。与此同时，自1955年美

注释

① ply [plaɪ] *v.* 定时往来，定期行驶
② diesel ['diːz(ə)l] *n.* 柴油，柴油车，内燃机车

thousands of nuclear-powered private machines to ply① the highways and byways of the land. Though perhaps they would if vehicles all travelled on an automated road, controlled by a computer; another idea that has been knocking around since the 1950s. It's easy to laugh at the notion, but, with oil production soon set to peak, how long does the combustion engine really have left? While the motor industry is working on cars powered by alternative fuel sources from electricity to hydrogen, the resulting machines have yet to make a significant dent in the consumption of petrol and diesel②.

In the future we may certainly need the sort of revolutionary idea embodied in the Nucleon, whose designers boldly stated in their 1958 brochure that they refused to 'admit that a thing cannot be done simply because it has not been done'.

国军舰**"鹦鹉螺号"**出海，核动力潜艇就问世了。

虽然核燃料的军事应用常常被视为足够安全，但是现在的政府绝对不可能允许成千上万的私家核动力汽车出现在大街小巷。不过，如果所有汽车行驶的道路都是由计算机控制的自动化公路——这是自20世纪50年代就出现的另一个想法，私人使用核动力汽车也许就会被允许。嘲笑这个想法很容易，但是，现在石油的产量很快就达到最大，内燃机还能用多久呢？汽车工业一直在研制如何在汽车上使用替代能源——从电力到氢能源，可至今仍未看到人们对汽油和柴油的消耗量出现明显减少。

将来，我们也许还需要体现在"核子号"上的那种革命性创意，它的设计师在1958年的宣传册上就大胆声明，他们绝不会"仅仅因为一件事从未做过，就承认不可能做成这件事"。

注释

"鹦鹉螺号"核潜艇，是美国海军的一艘核潜艇，是世界上第一艘核潜艇，也是第一艘从水下穿越北极的潜艇。"鹦鹉螺号"核潜艇的命名是为了纪念儒勒·凡尔纳小说《海底两万里》中的"鹦鹉螺号"潜艇。它于1952年6月14日在美国通用电船公司开工建造，1980年3月3日退役，之后经过改装在美国格罗顿潜艇部队作为博物馆艇。

ABANDONED
CINCINNATI'S SUBWAY TO NOWHERE

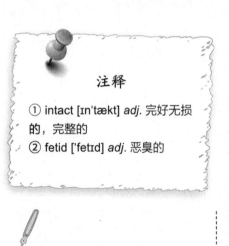

There are long platforms with huge flights of steps and miles of tunnels. But no trains ever arrive, there are no passengers to get on them and it's been that way for more than eighty years.

Many of the world's major conurbations have some form of underground railway, metro or subway. So does the city of Cincinnati in the US state of Ohio, with one big difference. Its subway system, much of it still intact① beneath the streets of the modern city, has never been used. As white elephants go, this is a monster. Far from being one of crowning glories of the Queen City, as Cincinnati is known, the subway has since been described as, 'one of the city's biggest embarrassments'.

The tale of this ghost network goes back to 1884 when the local *Graphic* newspaper suggested draining a fetid② canal that ran through the metropolis and replacing it with a subway. It wasn't until the early 1900s that a plan to make the subway a reality was produced and citizens of Cincinnati finally got their say on the idea in 1916. By then cities such as New York, and of course London, had underground or subway systems with some sections running above ground, others below. Cincinnati, the tenth largest city in the USA, was expanding fast and

辛辛那提不通车的地铁

长长的站台，巨大的阶梯，数千米长的隧道，但是没有火车，没有乘客，像这样已经80多年了。

世界上许多有卫星城的大都市都有地铁。美国俄亥俄州的辛辛那提也有地铁，但有一点大不相同——这座现代都市街道下面的地铁系统从未开通，许多地方甚至没被动过。对这座城市而言，这岂止是一个包袱，简直就是一座沉重的大山！它不是"皇后城"辛辛那提桂冠上的荣耀，而是"这座城市最大的尴尬"。

这个幽灵地铁网的故事始于1884年，那时当地报纸《图像》建议排干一条贯穿全市的沟渠，把它改造成地铁。20世纪初政府才制订建造地铁的计划，而在1916年辛辛那提的市民才在这个问题上有了发言权。那时，一些城市如纽约，当然还有伦敦，已经有了地铁系统，地铁部分建在地面上，部分在地底下。正在迅速扩张的美国第十大城市辛辛那提，也想拥有自己

注释

① poll [pəʊl] *n.* 选举投票，计票
② loop [lu:p] *v.* 使成环，使绕成圈
③ estimate ['estɪmeɪt] *n.*（对数量、成本等的）估计，估价
④ excavation [ekskə'veɪʃ(ə)n] *n.*（对古物的）发掘，挖掘
⑤ lucrative ['lu:krətɪv] *adj.* 赚大钱的，获利多的
⑥ succumb [sə'kʌm] *v.* 屈服，屈从，抵挡不住（攻击、疾病、诱惑等）
⑦ barrel ['bær(ə)l] *n.* 桶
⑧ tout [taʊt] *v.* 标榜，吹捧，吹嘘
⑨ burgeon ['bɜ:dʒ(ə)n] *v.* 激增，迅速发展
⑩ halt [hɔ:lt] *n.* 停止，阻止，暂停
⑪ boulevard ['bu:ləvɑ:d] *n.*（市区的）林荫大道

aimed to join the big players. Despite 200 miles of tram tracks, it desperately needed to solve its traffic problems too. So residents went to the polls① and overwhelmingly approved the construction of a 16-mile subway which would loop② round the centre of the city and its suburbs, at a cost of $6 million paid through an issue of bonds. A basic fare was even agreed – 5 cents a journey.

Then, crucially, in April 1917, America entered the First World War and work on Cincinnati's subway was put on hold. Once the war was over, local politicians decided to continue with their grand plan, even though inflation saw costs of steel and concrete soaring well above the plan's original estimates③. On 28 January 1920 excavation④ began. Completion of the project was expected in five years. Financial pressure soon meant that the original plan was reduced to 11 miles. But by 1923 the underground section of the subway had been finished. Costs kept going up. Yet so did the building work, thanks to the municipal authorities who still had a pot of cash and lucrative⑤ contracts to hand out. The city, which had been built on pork packing, was succumbing⑥ to pork barrel⑦ politics. By 1924 they were up to nearly 7 miles.

Then, in 1926, the new no-nonsense mayor, Murray Seasongood, revealed the bad news: another $10 million would be needed to save the subway. There was also a much more important player on the scene – the automobile. When the subway had first been touted⑧ there were few cars on the streets. But the decade of the 1920s saw the burgeoning⑨ of the American love affair with the car. Ownership of cars in the Cincinnati region went up by 126 per cent between 1921 and 1926, making the subway seem less essential.

In 1927 work on the subway was brought to a halt⑩. In 1928, the airy boulevard⑪ above, the Central Parkway, was opened to

的地铁。尽管已经有了约320千米（200英里）长的电车路轨，但辛辛那提的交通问题仍亟待解决。于是政府做了民意调查，市民们普遍支持围绕市中心和郊区修筑一条约26千米（16英里）长的地铁，工程造价为600万美元，通过发行债券集资。就连基本票价也都定好了——5美分一次车程。

接着发生了一件大事，1917年4月，美国参加第一次世界大战，辛辛那提地铁工程计划于是被束之高阁。战争一结束，当地政客决定继续实现这一宏伟蓝图，虽然通货膨胀已使钢铁和混凝土价格远远高出项目的最初预算。1920年1月28日，隧道挖掘工作开始了，整个项目预计5年完成。不久，财政上的压力使原计划缩减为约18千米（11英里）。到1923年，地铁的地下部分已经完工。成本在不断上涨，工程也仍在继续，这得感谢市政府的资金和一些赚钱的项目。辛辛那提是以猪肉加工为基础发展起来的城市，深受**"猪肉桶"**政治的影响。到1924年，约11千米（7英里）的地铁已经完工。

然后，在1926年，务实的新市长莫里·西曾古德（Murray Seasongood）宣布了一条坏消息：要挽救这条地铁，还需要1000万美元。这时，出现了一个更为重要的因素——汽车。当地铁最初引起公众注意的

▲ 20世纪20年代，美国辛辛那提市的地铁正在施工，这些隧道从未通过车。

The US city of Cincinnati's subway under construction in the 1920s. The tunnels would never be used.

注释

"猪肉桶"是美国政界经常使用的一个词。议员在国会制定预算时将钱拨给自己的州或自己热心的某个项目，这种做法叫作"猪肉桶"。

much celebration. Most of the revellers would soon forget that below their feet lay a near-complete subway network on which the city had blown its cash. What remained were miles of empty parallel① tunnels, with three complete stations situated in the section that ran underground at Liberty Street, Race Street and Brighton's Corner, all now concealed beneath the hustle② and bustle③ of the city's roads. There were also three stations above ground ready for action at Marshall Street, Ludlow Avenue, and Clifton Avenue, which were later demolished④, to make way for more roads. Once construction had stopped hopes of getting the project back on its feet were totally dashed following the 1929 Wall Street Crash and subsequent Great Depression. A line was finally drawn under the project in 1948.

In the decades that followed, debate has raged⑤ about what to do with the empty subway system. Proposals to use the tunnels as roads or for new rail systems faltered⑥. During the Cold War a part of the system was even turned into a nuclear fallout⑦ shelter, also since abandoned. A massive wine cellar was once mooted⑧ and the atmospheric tunnels were reportedly even considered as sets for the Hollywood blockbuster⑨ Batman Forever, but even the movie moguls⑩ pulled the plug⑪.

So there they still sit, several miles of defunct tunnels and stations minus the steel rails and trains that will never come; a monument to the financial and political perils⑫ of town planning. It took until 1966 for Cincinnati to pay off the original bonds for the subway at an extra cost of around $7 million. The administration has since shelled out more cash to keep the subway mothballed⑬, partly to make sure the road above doesn't fall in. The alternative is spending huge sums to fill in the tunnels.

Cincinnati might be the best example of an underground system which was built but never used. But other networks

时候，街上还没有多少汽车。但到了20年代，美国人开始爱上了汽车。1921—1926年，辛辛那提地区拥有汽车的人数增加了126%，地铁显得没那么重要了。

1927年，地铁停工了。1928年，地面上通风的林荫大道、中央公园大道在盛大的庆典中开通了。许多狂欢者很快忘记了他们脚下还躺着一个几近完工的地铁网络，以及这座城市为它烧了多少钱。最后这个工程只留下几条数千米长的并排空隧道和三个完工的车站，分别位于自由街、礼士街和布莱顿角，现在都被隐藏在城市喧嚣的路面下。

另外，地面上还有三个车站已准备好投入使用，分别位于马歇尔街、勒德洛大道和克利夫顿大道，后来因为修路被拆掉了。项目停工后，1929年又发生了华尔街股市崩盘以及之后的经济大萧条，重新启动这个项目的希望完全破灭了。1948年，这个工程寿终正寝。

在后来的几十年里，人们对如何处理这个只有空架子的地铁系统进行了激烈讨论。有人提议将隧道作为公路使用，或建成新的铁路系统，但未能站得住脚。在"冷战"期间，这个地铁系统的一部分被改建为核放射性尘埃掩蔽所，后来也被废弃了。人们还讨论过这里是否可以用作一个巨大的酒窖。据报道，通风隧道还被考虑作为好莱坞大片《永远的蝙蝠侠》的摄制外景，但即便是电影大腕也无法挽救这个地铁的命运。

就这样，它们静静地躺在那里，数千米废弃的隧道和车站，没有铁轨，没有列车，成为城市规划经济和政治风险的警示牌。到1966年，辛辛那提市才还清地铁修建初期发行的债券，多支出的成本高达700万美元。管理部门后

① ghostly ['gəʊs(t)lɪ] *adj.* 幽灵的，可怕的，影子似的
② deem [diːm] *vt.* 认为，视作
③ millstone ['mɪlstəʊn] *n.* 磨石，重担
④ haunt [hɔːnt] *vt.* 常出没于……，萦绕于……
⑤ disgruntled [dɪs'grʌnt(ə)ld] *adj.* 不满的，不高兴的

have their ghosts too, including London's underground. If you look carefully in the suburbs you can still see the remnants of the extension to the Northern Line that was partially built in the 1930s. The Northern Heights scheme would have seen the line extended to Bushey Heath with stations at Elstree South and Brockley Hill. A route was laid out and bridges built. You can even see some of the arches put up to carry trains into the Bushey Heath Station that never was. But once again a war intervened – the Second World War. Afterwards, new green belt planning restrictions meant that the area wasn't going to develop as had once been expected, so there just wouldn't be enough passengers for the line.

Around the world there are more ghostly① stations in subway systems. Paris has a station in its network called Haxo, completed but never used as it was eventually deemed② unprofitable. Stockholm has its own example, Kymlinge. Yet none of these has ever been the millstone③ that the Cincinnati subway became for city officials. It's hardly surprising that tales emerged of the tunnels being haunted④ by the spirits of construction workers who were unlucky enough to die during the building of their transit system. You could forgive them for feeling a little disgruntled⑤. After all, its demise wasn't their fault. As one Cincinnati council member summed up the subway in 2007: 'It didn't go anywhere, but it was built well.'

来还花了更多的钱来维护这个地铁，部分工作是确保路面不塌陷。否则，就要花巨资把隧道填起来。

辛辛那提地铁或许是建造地铁失败的典型案例。不过，其他地铁网也有它们自己的"幽灵路段"，包括伦敦地铁。如果你仔细观察伦敦郊区，就能发现北线延长部分的地铁残余，那里部分是20世纪30年代修建的。据北部高地计划，这条地铁本来要延伸至布西希思，还会在埃尔斯特里南和布洛克利山修建车站。地铁线布置好了，地铁桥也建好了。甚至今天你都能看到通往那个未建成的布西希思地铁站的一些立起来的拱顶。但是，又是战争耽误了工程，这次是第二次世界大战。后来，政府制订了新的绿化带规划，这意味着这个地区不会像当初预计的那样发展，也不会有足够的乘客来支撑这条线路。

世界上还有许多"幽灵地铁站"。巴黎的哈克索车站虽已建好，但从未使用过，因为人们认为它不赚钱。斯德哥尔摩也有这样的地铁站——科姆林吉车站。但是，没有哪个地铁会像辛辛那提的地铁一样成为市政官员的重担。民间流传着一些故事，说那些修建地铁时不幸死去的建筑工人的鬼魂在隧道出没。这不足为奇，不过，就算这些鬼魂有些不满，你也应该谅解。毕竟，你不能把地铁的失败怪到他们头上。一位辛辛那提议员在2007年这样概括这条地铁："它没有用，但建得很好。"

~~CANCELLED~~
IS IT A TRAIN OR A PLANE?

In the 1920s railways still ruled when it came to mass transport over big distances. And, with some rare exceptions, steam power still ruled the world's railways. Yet inventors like George Bennie were looking for new ideas to revolutionise rail travel. Some were experimenting with diesel① designs, others with electric units. Bennie, the son of a Glaswegian hydraulics② engineer, had an idea to combine the sleek③ lines and technology employed by the age's new aircraft with what he saw as the reliability of a land-based transport system. And so, in 1921, the Bennie Railplane was born.

This was a far cry from some of Bennie's other inventions. His improved golf putter④ for example, which featured a special boss on the face of the club for extra accuracy, had not proved eye-catching. His futuristic, streamlined Railplane, on the other hand, provided a startling spectacle. And, on 8 July 1930 when the gleaming⑤ prototype was shown off to the press and potential investors, Bennie boasted⑥ that his Railplane would be capable of 120 mph. This at a time when express steam trains were only averaging 80 mph.

注释

① diesel ['diːz(ə)l] n. 柴油机，柴油

② hydraulics [haɪ'drɔːlɪks] n. 水力学

③ sleek [sliːk] adj. 光滑的，光亮的

④ putter ['pʌtə] n. 轻击球杆

⑤ gleaming ['ɡlimɪŋ] adj. 闪耀的，明亮的

⑥ boast [bəʊst] v. 自夸，自吹自擂

它是火车还是飞机？

　　20世纪20年代，铁路在大规模、远距离运输方面占据着统治地位，而除极少数例外，蒸汽动力则在铁路运输中占据着统治地位。不过乔治·本尼（George Bennie）等发明家正在寻求改革轨道交通的新想法，一些人在试验使用柴油机车，另一些人则在试验使用电力机车。本尼是格拉斯哥一位水力工程师的儿子，他萌发了把当时飞机所用的流线外形技术与铁路这种可靠的地面交通系统相结合的想法。于是在1921年，本尼的铁路飞机诞生了。

　　这与本尼的其他发明迥然不同。例如，他改进了高尔夫球推杆，其特色是在球棒表面上具有特别的突起，可以提供额外的准确性，但效果并不十分引人注目。然而另一方面，他设计的流线型铁路飞机十分新潮，极大地提高了效率，令人耳目一新。1930年7月8日，他向媒体和潜在投资人展示了闪亮的"样机"，声称他的铁路飞机速度可以达到约193千米/时（120英里/时），当时高速蒸汽火车的平均速度仅为约129千米/时（80英里/时）。

The Railplane looked a bit like the kind of monorail① used in some cities today. But the cars were driven by four bladed propellers② at either end powered by either live electric rails or combustion engines. These cars were suspended from a large metal gantry③ with a guide rail at the bottom. Bennie intended that hundreds of miles of this gantry would be constructed across the country above existing railways. This would free the traditional railways up for freight, easing congestion④. Meanwhile passengers would be hurtled to their destinations at lightning speeds in modern comfort high above the tracks.

Bennie claimed the Railplane would be economical to build too. He estimated the construction cost of an ordinary railway at £47,500 a mile, while his Railplane would cost just £19,000 a mile. Most importantly it would be almost as fast as going by aircraft. At the time these were pretty ponderous⑤, had limited capacity and were often grounded in bad weather. Bennie believed his Railplane would be safer than taking to the skies yet be able to whisk travellers from, say, Glasgow to Edinburgh in just twenty minutes and from Glasgow to London in three and a half hours.

After years of finessing⑥ the design, with the help of an engineer called Hugh Fraser, Bennie persuaded one of the biggest rail companies in the land, the London and North Eastern Railway, to let his company construct a test line over a rail siding at Milngavie in East Dunbartonshire. The 426ft line was 80ft across and 16ft from the ground. His prototype car was 52ft long, 8ft in diameter⑦, weighed in at 6 tons and was designed to take fifty passengers. The look of the thing owed a lot to the airships of the time and indeed its engineers, William Beardmore and Co, had built the R34 airship, which made the first airborne, east to west, Atlantic crossing in 1919.

这种铁路飞机看起来有点像现在一些城市用的单轨列车。但这种火车是靠带电电轨或内燃机在任一端驱动的四个螺旋桨推动的，把车身悬挂在巨大的金属架上，下面悬有导轨。本尼提议在现有铁路上方建造数百千米金属架，这样就可以将传统铁路更多地用于货运，有效缓解交通拥堵。同时，乘客将以闪电般的速度在轨道上方舒适地到达目的地。

本尼声称，铁路飞机建造成本很经济。他估计普通铁路的建设成本约为29516英镑/千米（47500英镑/英里），而铁路飞机只需要约11806英镑/千米（19000英镑/英里）。最重要的是，乘铁路飞机几乎和乘飞机一样快。当时的飞机十分笨重，载重有限，而且经常在恶劣天气停飞。本尼认为，乘铁路飞机比乘飞机更为安全，而且也能在短短20分钟之内使乘客从格拉斯哥到达爱丁堡，在3.5个小时之内从格拉斯哥到达伦敦。

之后几年，在工程师休·弗雷泽（Hugh Fraser）的帮助下，本尼不断改进设计方案。后来，他说服英国最大的铁路公司之一，伦敦和东北铁路公司，让这家公司在东邓巴顿郡米尔盖的铁路上方建了一条试验线路。这条线路长约130米（426英尺），跨度约24米（80英尺），离地约5米（16英尺）。他的样车长约16米（52英尺），直径约2.4米（8英尺），重6吨，设计载客50人。样车的外观与当时的飞船极

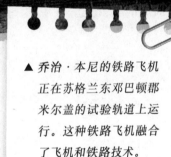

▲ 乔治·本尼的铁路飞机正在苏格兰东邓巴顿郡米尔盖的试验轨道上运行。这种铁路飞机融合了飞机和铁路技术。

George Bennie's Railplane prototype running on a test track in Milngavie, East Dubartonshire, Scotland. The Railplane was a fusion of aircraft and railway technology.

注释

① plush [plʌʃ] *adj.* 舒适的，豪华的
② interior [ɪn'tɪɜːrɪə] *n.* 内部，里面
③ reverse [rɪ'vɜːs] *adj.* 相反的，反面的，反向的
④ thrust [θrʌst] *n.* 推力，驱动力
⑤ hark [hɑːk] *v.* （用于命令）听着，听
⑥ luminary ['luːmɪn(ə)rɪ] *n.* 专家，权威，有影响的人物
⑦ genuine ['dʒenjuɪn] *adj.* 真的，名副其实的
⑧ buffer ['bʌfə] *n.* 缓冲物，起缓冲作用的人
⑨ bankrupt ['bæŋkrʌpt] *n.* 破产，倒闭
⑩ sideline ['saɪdlaɪn] *n.* 兼职，副业，兼营业务

Newsreels of the Railplane's launch show plush① interiors② with comfortable armchairs and a saloon. The sliding doors even featured elaborate stained glass. Technologically the Railplane was well thought out. As a train approached the station the front propeller would stop while the back one was switched on to achieve reverse③ thrust④, with brake shoes gripping the rails. But one of the most positive aspects would be the lightweight nature of the whole design that meant there was little track resistance. In a brochure for his Railplane, which harks⑤ back to the glorious rail breakthroughs by the likes of George Stephenson, Bennie also points out that the Railplane can tackle hills easily.

Things looked positive. One reporter described travelling on the Railplane in a test run as a 'sheer delight' and luminaries⑥ such as prime minister Ramsay MacDonald took an interest. Council officers in Blackpool, Lancashire were keen to investigate how the Railplane could be used for a link to nearby Southport, and there was even genuine⑦ interest from Palestine.

Sadly, like many good ideas, the timing was off. The Great Depression meant money was tight and, in the end, no orders came in quickly. Perhaps Bennie was too ambitious. He made a submission to the government in 1935 to build a line from London to Paris involving a seaplane link across the channel. Bennie reckoned that his link would speed people between the capitals in 140 minutes. A contemporary aircraft would take 225 minutes. This proposal hit the bureaucratic buffers⑧ too. So convinced had he been in success that Bennie had already invested £150,000 of his own money in the project. By 1937 he was declared bankrupt⑨ and with the Second World War on the horizon, revolutionary schemes like Bennie's were destined to end up sidelined⑩.

像。事实上，样车的设计师威廉·比尔德莫尔（William Beardmore）公司曾经建造了R34飞船，这是1919年首次从东至西飞越大西洋的飞行器。

铁路飞机试运行的新闻短片展示了车内的豪华装饰，里面有舒适的沙发和餐厅，推拉门上甚至还装饰了精美的彩色玻璃。铁路飞机从技术上来说设计十分完美。当铁路飞机快到站时，前螺旋桨停止工作，后螺旋桨启动，从而产生后向推力，刹车片抓住铁轨。不过最有利的一个特征还是整个设计的重量极轻，所以产生的轨道阻力很小。在铁路飞机宣传手册中，本尼还指出铁路飞机可以轻快地翻越崇山峻岭，这让人们又回想起乔治·斯蒂芬逊（George Stephenson）等人在铁路方面所创造的光辉突破。

事态进展似乎非常顺利。一位记者说，在一次试运行中乘坐铁路飞机飞行"十分愉快"，而像首相拉姆齐·麦克唐纳（Ramsay MacDonald）等名人也对此表示很感兴趣。兰开夏郡黑泽市议会官员非常热衷于勘察一下如何采用铁路飞机与附近的南波特连接起来，甚至巴勒斯坦也有人表示对此很感兴趣。

不幸的是，铁路飞机像许多好想法一样生不逢时。大萧条时期资金十分紧张，结果没有迅速接到订单。也许本尼太雄心勃勃了，他在1935年向政府提议建设一条从伦敦到巴黎的铁路线，采用水上飞机穿越海峡将两国相连。本尼估计这条铁路线可以使人们在140分钟内穿梭于两国的首都之间，而当时的飞机尚且需要225分钟。这项提案也没有打动那些官僚作风十足的老糊涂们。本尼相信自己一定能够成功，早就把自己的15万英镑倾囊投入这个项目。

There had been other attempts to fuse[①] aircraft and rail technology abroad, with equally unconvincing results. A prototype Russian rail plane called the Aerowagon, powered by an aircraft engine and propeller, derailed[②] while being tested in 1921; killing everyone on board including its inventor, Valerian Abakovsky. Franz Kruckenberg had more success launching his 'rail zeppelin in 1929; it even reached an impressive 160 mph. Technical issues, however, including the fact that the era's tracks really couldn't take a train going that fast, meant it was a commercial non-starter.

Despite attempts to re-boot the Railplane after the war – Bennie even took his ideas to the Iraqis – the world had moved on and the remains of his Railplane prototype and its test track were finally demolished[③] for scrap[④] in 1956. Bennie himself died a year later aged just 65, having, by some accounts, given up designing trains to become a herbalist[⑤]. His design did have its flaws. It was noisy and his plan for junctions involved time consuming turntables[⑥]. Doubts remain about its viability[⑦] for long-distance travel, which was Bennie's real dream, and surely its unsightly[⑧] gantry[⑨] would have thrown up plenty of planning issues. It was probably more costly to build than Bennie had calculated. Nor did he foresee technological improvements by planes or the growing expansion of road travel.

In comparison to some other railway innovators of the time, however,

注释

① fuse [fju:z] v. （使）融合，熔接，结合
② derail [dɪ'reɪl] v. （使）脱轨，出轨
③ demolish [dɪ'mɒlɪʃ] v. 拆毁，拆除（建筑物）
④ scrap [skræp] n. 碎片，小块（纸、织物等）
⑤ herbalist ['hɜ:b(ə)lɪst] n. 药草栽培者，药草商，草药医生
⑥ turntable ['tɜ:nteɪb(ə)l] n. （铁路机车的）转台，旋车盘
⑦ viability [ˌvaɪə'bɪlətɪ] n. 生存能力，发育能力，可行性
⑧ unsightly [ʌn'saɪtlɪ] adj. 难看的，不雅观的
⑨ gantry ['gæntrɪ] n. （起重的）龙门架

到了1937年，他宣告破产，而第二次世界大战的爆发也注定了像本尼的铁路飞机一样具有革命性的方案走向了穷途末路。

还有其他国家的人曾尝试把飞机与铁路技术结合起来，但结果同样不尽如人意。俄罗斯人设计的一架被称为"空中马车"的铁路飞机，以飞机发动机和螺旋桨为动力，在1921年的试验中脱轨，乘客全部遇难，包括发明者瓦莱利安·阿巴科夫斯基（Valerian Abakovsky）。弗兰茨·克鲁肯伯格（Franz Kruckenberg）于1929年推出了更为成功的铁路飞艇，时速竟然达到了约258千米/时（160英里/时）。然而，当时的轨道无法承载速度那么快的火车，同时还存在其他一系列等技术问题，使铁路飞艇也在商业化过程中夭折了。

尽管本尼在战后曾尝试重启铁路飞机项目，甚至把这一想法带到伊拉克，但是世界在发展，他的铁路飞机样机的残骸和试验轨道最终于1956年被拆除。他本人也于一年之后去世，年仅65岁。在这之前几年他就放弃了设计火车，转而成为一位药材商人。他的设计确实存在缺陷，不仅噪声大，而且交叉方案涉及耗时的转换。本尼的真正梦想其实是实现长途交通，而人们对长途交通的可行性仍然疑虑重重。而难看的金属架又涉及大量规划问题，这可能比本尼估算的成本要高出很多。此外，他也没有预测到飞机的技术改进和公路交通的快速扩展问题。

然而，与当时的其他铁路革新者相比，本尼的想法

注释

① cautious ['kɔːʃəs] *adj.* 谨慎的，十分小心的
② smithereen [ˌsmɪðə'riːn] *vt.* 把……击碎，把……炸成碎片

Bennie's ideas look positively cautious①. How about a train powered by rockets? Sounds like a one-way ticket to disaster, but Fritz von Opel, who invented it, thought otherwise. He was grandson of the Adam Opel, the founder of the famous car company, and in 1928 he tested a rocket-powered car, the Rak I. Later the same year he decided to put his rocket car on rails. On a closed set of tracks his Rak 3 reached the amazing speed of 157 mph. Attempting to beat the record a few weeks later his new improved Rak 4 exploded into smithereens②, effectively putting an end to any similar experiments.

还算谨慎的。火箭动力火车怎么样呢？听起来就像是通往灾难的单程票，但它的发明者弗里茨·冯·奥佩尔（Fritz von Opel）却认为并非如此。他是著名汽车公司欧宝的创始人亚当·奥佩尔（Adam Opel）的孙子，在1928年试验了一种火箭动力汽车Rak-1号。1928年年底，奥佩尔又决定把火箭动力汽车放在轨道上。在一套闭合轨道上，他的Rak 3号达到了约253千米/时（157英里/时）的惊人速度。为了打破几周前创下的纪录，他的改进型Rak 4炸得粉碎，彻底结束了之后的类似实验。

ABANDONED
HOW THE CAPE TO CAIRO
RAILWAY HIT THE BUFFERS

注释

① colonialist [kə'ləʊnjəlɪst] *n.* 殖民主义者
② unflinching [ʌn'flɪn(t)ʃɪŋ] *adj.* 不畏缩的，不退缩的
③ formidable ['fɔːmɪdəbl] *adj.* 可怕的，难对付的

In 1877 when British-born Cecil Rhodes was 24 years old and studying as an undergraduate at Oxford, he made an extraordinary statement as part of his will. It read: 'I contend that we are the first race in the world and that the more of the world we inhabit the better it is for the human race.' Brought up in Southern Africa, Rhodes was an unashamed colonialist① who helped the British Empire expand its territories in the southern part of the continent in the latter part of the nineteenth century. He gave his name to the colonies of Southern and Northern Rhodesia, now Zimbabwe and Zambia, and also set up the famous diamond company De Beers. You'd have thought that would be enough, but not for Rhodes. It was an unflinching② belief in the superiority of British rule, an iron will and a canny entrepreneurial brain that led to Rhodes' idea to link up British territories across Africa, from the very north to its southernmost tip. The backbone of this plan would be the Cape to Cairo railway, a line more than 5,000 miles long which would have to cross some of the most remote, most disputed and most physically challenging territory anywhere in the world.

The odds against such a railway were always formidable③.

好望角至开罗的铁路
如何遭遇搁浅

1877年，出生于英国的赛西尔·罗得斯（Cecil Rhodes）时年24岁，正在牛津大学攻读学士学位，他在他的遗嘱中说了一番超乎寻常的话。他说："我认为我们是世界上最上等的种族，我们在世界上居住的面积越大，对人类越有利。"罗得斯在南非长大，是一个不知羞耻的殖民主义者，帮助大英帝国在19世纪后半叶将领土扩张到了非洲大陆的南部。他用自己的名字将两块殖民地命名为南罗得西亚和北罗得西亚，也就是今天的津巴布韦和赞比亚，而且还成立了著名的钻石公司戴比尔斯。你可能认为这已经足够了，但对罗得斯来说还远远不够。正是对英国统治至高无上的坚定信念、钢铁般的意志和精明的头脑，使罗得斯产生了把英国在非洲的领地从最北端至最南端连接起来的想法。这一计划的主干是好望角至开罗的铁路，这条铁路长8000多千米（500多英里），将必须穿过世界上最偏远、最具争议、最具挑战性的领土。

In Russia the Trans-Siberian Railway, of a similar distance, was well under way by the turn of the nineteenth century. But Rhodes' railway would have to be built in a climate just as punishing and its workers would have to contend with hostile① locals and dangerous animals, all against the backdrop② of political turmoil③ as the great powers scrambled④ to conquer the remaining parts of the continent.

In the 1890s Rhodes certainly looked like a man that could make it happen. He had become prime minister of the British held Cape Colony in the very south and developed a serious reputation as an Empire builder by securing new territories for the British crown to the north. These became known as Rhodesia, under the auspices of a company he'd set up, the British South Africa Company. A cartoon which appeared in the satirical *Punch* magazine in 1892 was titled The *Rhodes Colossus* and portrayed him as a giant astride⑤ the whole continent.

In the same period, Rhodes furthered the construction of railways in Southern Africa and raised much of the finance himself. By 1897 a railway had got as far as Bulawayo, Rhodesia, 1,360 miles from Cape Town. The moment was celebrated with great pomp⑥ and Queen Victoria even sent a telegram of congratulations. This achievement was largely down to the efforts of a thrusting⑦ railway engineer, George Pauling, who pushed the track hundreds of miles through the inhospitable⑧ Kalahari Desert at breakneck⑨ speed. In one day alone 8 miles of track had been laid. Rhodes, suffering from fever, wasn't at the Bulawayo celebrations, but also sent a telegram vowing⑩ to forge the railway onwards to Victoria Falls.

At the other end of the continent there was another breakthrough.In September 1898, British forces under the

不利于这条铁路的因素一直难以克服。19世纪末，俄罗斯也建了一条类似距离的西伯利亚大铁路，但罗得斯的铁路也必须在类似的严峻环境下建设，筑路工人必须与充满敌意的当地人和危险的动物作斗争，加之当时各大强国都在争相霸占非洲大陆的其他地方，因而造成了这一地区严重的政治动荡。

19世纪90年代，罗得斯似乎可以使这条铁路成为事实了，他成了非洲南部英控好望角殖民地的首相，通过向北部拓展领土而荣获"帝国建设者"的殊荣。这些领土被命名为"罗得西亚"，处于他所成立的英国南非公司的统治之下。1892年，讽刺类杂志《笨拙》上出现了一幅名为"巨人罗得斯"的漫画，把罗得斯描绘成一位横跨非洲大陆的巨人。

同时，罗得斯又筹集了大量资金，进一步加强南非铁路的建设。到1897年，铁路已经抵达罗得西亚的布拉瓦约，距开普敦约2190千米（1360英里）。人们隆重庆祝了这一时刻，维多利亚女王甚至也发来贺电。这一成就在很大程度上取决于一位自负的铁路工程师乔治·鲍林（George Pauling）。他以极快的速度在寸草不生的卡拉哈里沙漠铺设了数百千米铁轨，曾经在一天内就铺了约13千米（8英里）。罗得斯身患伤寒，未能出席布拉瓦约铁路的通车庆典，但也发

▲ 巨人罗得斯，1892年《笨拙》杂志上的一幅漫画，将赛西尔·罗得斯描绘成一位横跨非洲的巨人。

The Rhodes Colossus, an 1892 cartoon from Punch magazine showing empire-builder Cecil Rhodes astride Africa.

注释

① tussle ['tʌs(ə)l] *n*. 扭打，争斗，争执（尤指为了争得物品）
② ferry ['ferɪ] *n*. 渡船，摆渡
③ annex [ə'neks] *v*. 强占，吞并（国家、地区等）
④ torch [tɔ:tʃ] *n*. 火炬，火把

command of Lord Kitchener secured control over Northern Sudan at the Battle of Omdurman, where he defeated the local Mahdi's forces. And the same month, a tussle① with the French in what was called the Fashoda Incident secured the south of the Sudan, putting an end to French hopes of building their own railway east to west across Africa. This enabled a railway and ferry② link to extend 1,000 miles south from Cairo to Khartoum in Sudan, completed in 1899.Following the victories Rhodes had even cheekily cabled Kitchener, saying: 'If you don't look sharp … I shall reach Uganda before you.' Uganda, in the middle of the continent, had recently come under British rule too. Kitchener swiftly sent a telegram back saying: 'Hurry up!'

By the time Rhodes died in 1902, aged just 49, the Boer War had seen Britain annex③ the independent Boer republics, leading ultimately to the Union of South Africa and the consolidation of British territory in the region. Rhodes may have gone, but Pauling carried the torch④ onward. In 1904 the section extending northwards with a spectacular bridge to Victoria Falls was completed, allowing tourists easy access to the natural wonder for the first time. The project was another 300 miles towards its goal and the Cape to Cairo dream still seemed very much alive. But there was a big problem ahead – the Germans. They controlled German East Africa, modern day Tanzania. While the Kaiser did allow Rhodes to build a telegraph line through his territory, he wasn't going to let the British connect up their railway lines through it.

George Pauling continued to push north nevertheless, building the line to the border with Belgian-controlled Congo instead. This came with the added temptation of opening up new mineral mines. In 1908 the *New York* Times reported that there were only 700 miles to go before it would be possible to make the trip by lake, river and railway from Cairo to Cape

来电报，发誓要把这条铁路修到维多利亚大瀑布。

在非洲大陆的另一端，又有了另一个重大突破。1898年9月，在基切纳勋爵的指挥下，英国军队在乌姆杜尔曼战斗中打败当地的马赫迪起义者，占领了苏丹北部。同月，在名为"法绍达事件"的与法国的争端中，英国又占领了苏丹南部，使法国修建一条横跨非洲的铁路的希望破灭了。这就使得铁路和轮渡线从开罗向南延伸了1600多千米（1000英里），达到苏丹的喀土穆，最终于1899年竣工。在这一连串的胜利之后，罗得斯甚至愉快地给基切纳发报说："如果你再不快点儿，我就要在你之前到达乌干达了。"乌干达位于非洲大陆中央，很快也处于英国的统治之下。基切纳立即回报说："快点儿！"

1902年罗得斯去世时，年仅49岁。这时布尔战争爆发，英国吞并了布尔人的共和国，最终导致南非联邦和英国在这一地区领土的统一。罗得斯已经走了，但鲍林挑起重担，继续前进。1904年，铁路继续向北延伸，落成了一座直达维多利亚大瀑布的大桥，使游客首次可以方便地到达这一自然奇观。这一项目离目标又近了约483千米（300英里），而从好望角至开罗的梦想仍然看起来充满活力。但还面临着一个重大问题——德国人。他们控制着德属东非，即今天的坦桑尼亚。虽然德国皇帝让罗得斯经过他的领地修建一条电报线路，但他不打算让英国人的铁路通过他的领地。

乔治·鲍林继续向北推进，把铁路修到了比利时控制下的刚果的边界，这就加大了开采新矿场的诱惑力。

注释

① navigable ['nævɪɡəb(ə)l] adj. 可航行的，适于通航的

② unadulterated [ʌnə'dʌltəreɪtɪd] adj. 完全的，十足的，不折不扣的

③ stretch [stretʃ] v. 拉紧，拉直，绷紧

④ arduous ['ɑːdjʊəs] adj. 艰苦的，艰难的

⑤ impetus ['ɪmpɪtəs] n. 动力，推动，促进，刺激

⑥ perilous ['perɪləs] adj. 危险的，艰险的

⑦ zealot <英> ['zelət] n. （尤指宗教或政治的）狂热分子，狂热者

⑧ tentative ['tentətɪv] adj. 不确定的，不肯定的，暂定的

Town. And, in December 1917, newspapers announced that the railway had reached Bukama on the banks of the navigable① Upper Congo where a steamer link could take travellers further north to Lake Tangynika. A person could now travel by train 2,600 miles north of Cape Town on the Congo Express in just six days. The defeat of Germany in the First Word War then seemingly gave Rhodes' vision an extra boost as German East Africa fell into British hands. Rhodes' idea of a continuous line of unadulterated② British territory stretching③ from the Cape to Cairo had come true.

At the same time his railway seemed to be dropping off the priority list. By the 1930s it was theoretically possible to make the journey overland using a series of railways, ferries and buses entirely through British held territory. A few more sections of line had been built in Uganda and Kenya but a line was never pushed through between Uganda and Sudan, and few were prepared to undertake the arduous④ journey. It seems that much of the political and financial impetus⑤ for the Cape to Cairo line had run out. In the interwar period Britain was preoccupied with perilous⑥ international relations as well as economic instability. The railways that did exist in Britishruled Africa had largely been built with private money and, with a global downturn, along with the absence of colonial zealots⑦ like Rhodes or Kitchener (killed at sea in 1916), there was little hope of pushing such a grandiose project through. With the eventual break-up of the Empire after the Second World War, a cross-continental railway dropped off the agenda completely. The new, independent African republics had much more pressing problems.

In recent years there have been tentative⑧ signs that the idea of a Cape to Cairo rail link could be reborn, though not, as Rhodes had hoped, overseen by the British. In the 1970s

1908年，《纽约时报》报道，修通一条从开罗至好望角的旅行路线，现在只差约1127千米（700英里）了。1917年12月，报纸宣布铁路已经修至适于航行的刚果河上游河岸的布卡马，从这里再用汽船把旅客向北送至坦噶尼喀湖。现在人们可以乘坐刚果特快在短短6天内到达开普敦以北约4184千米（2600英里）处。德国在第一次世界大战中的失败似乎给罗得斯的愿望注入了强劲动力，因为德属东非已经落入英国手中。罗得斯的那个让英国的领地从好望角到开罗连成一线，中间不夹杂任何其他国家殖民地的想法成了现实。

与此同时，罗得斯的铁路似乎已经不再是当务之急了。到了20世纪30年代，采用火车、轮渡和汽车完全通过英控领地进行陆地旅行在理论上有了可能性。英国又在乌干达和肯尼亚建了一段铁路，但乌干达和苏丹之间却始终没有进展，而且也没人愿意承担重任。似乎好望角至开罗铁路的政治和经济动力已经丧失了。在两次战争之间，英国深陷危险的国际关系和动荡的经济局势之中。英治非洲内存在的铁路主要是靠私人资金修筑的，随着全球大萧条的到来，加之殖民狂热者罗得斯和基切纳（1916年死于海上）等人已经去世，要把这么宏伟的项目推行下去，希望似乎十分渺茫。第二次世界大战之后，随着大英帝国的不断瓦解，跨洲铁路完全退出了议事日程。新独立的非洲共和国还有许多更紧迫的事情要做。

近年来，好望角至开罗铁路线的想法又有了复燃

注释

① afoot [ə'fʊt] *adj.* 计划中的，进行中的

the Chinese helped build a brand new railway line between Tanzania and Zambia which connected to the railways running down to the Cape. And, as of 2010, plans are afoot① to connect up the line in Uganda with the one in the Sudan. It is to be largely funded by a German company, a fact which would surely have Rhodes turning in his grave.

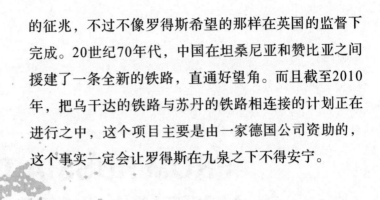

的征兆，不过不像罗得斯希望的那样在英国的监督下完成。20世纪70年代，中国在坦桑尼亚和赞比亚之间援建了一条全新的铁路，直通好望角。而且截至2010年，把乌干达的铁路与苏丹的铁路相连接的计划正在进行之中，这个项目主要是由一家德国公司资助的，这个事实一定会让罗得斯在九泉之下不得安宁。

REJECTED

FROM RUSSIA TO AMERICA BY TRAIN

At the turn of the twentieth century, in the same era that Rhodes was trying to connect up railways in Africa, another even more audacious① project was being considered – a transport link that would connect five continents and allow people to travel from Europe to America without ever having to take to the perilous high seas.

The key to this idea is the 53-mile-wide expanse of water known as the Bering Strait just south of the Arctic Circle. While remote and suffering from an inhospitable climate, this narrow gap at the very top of the Pacific Ocean is all that separates Russia on one side and the USA on the other. The first to identify the possibility of linking Russia and America via a railway was William Gilpin, a bombastic② ex-soldier and the first governor of the US state of Colorado. He was one of the nineteenth century's most pugnacious③ nation builders. In 1846 he had said: 'The untransacted④ destiny of the American people is to subdue⑤ the continent – to rush over this vast field to the Pacific Ocean … and shed⑥ blessings around the world.'

In Gilpin's lifetime railways helped open up North America's wild west to trade and migration, and he felt that

坐火车从俄国到美国

20世纪初，就在罗得斯尝试把非洲的铁路连接起来时，有人正在考虑另一个更为大胆的项目——建设一个连接五大洲的交通运输网，让人们不用再远涉重洋就能游历欧美。

这一设想的关键是位于北极圈正南部宽约85千米（53英里）的**白令海峡**。这条窄窄的海峡位于太平洋边缘，地势偏远、气候恶劣，使俄罗斯与美国隔海相望。第一位指出通过铁路可以把美俄相连的人是威廉·吉尔平（William Gilpin），他是一位夸夸其谈的退伍军人，是美国科罗拉多州的首任州长，是19世纪最好战的建国者之一。1846年，他曾说过："美国人民无法摆脱的宿命就是征服大陆——就是冲过广袤的太平洋……把福祉洒向全世界。"

在吉尔平生活的时代，铁路促进了北美洲荒凉西部的开发，给那里带去了贸易和移民，所以他认为铁路可以促进"新世界"与"旧世界"之间的联系。1890

注释

白令海峡位于亚欧大陆最东点的迭日涅夫角和美洲大陆最西点的威尔士王子角之间，连接楚科奇海和白令海，最窄处约37千米，深度30~50米。

trains could help connect the 'new world' to the 'old world' too. In 1890 he wrote a book called *The Cosmopolitan Railway: Compacting and Fusing Together All the World's Continents*. In it Gilpin spelled out how a rail route could be forged across America to Asia and Europe, making long journeys by ship across the Atlantic unnecessary. Thus one could 'transcend[①] the disharmony of world geography' and 'so bring together and intermingle[②] all the people of the earth as ultimately, in a great measure, to obliterate[③] race distinctions and bring about a universal brotherhood of man'.

First a railroad would be built across the icy wastes of Alaska, the US territory which had been purchased from the Russians in 1867, to the Bering Strait (named after Danish explorer Vitus Bering who crossed it in 1728). A similar railway on the other side would bring travellers to the Russian coast at Cape Dezhnev, the most Eastern point of the Asian continent. Rail passengers would then be able to transfer across the Strait on a rail ferry. Eventually this would allow travel by rail between Asia, Europe, Africa and both North and South America. Being a good Coloradan, Gilpin saw his own state's capital, Denver, being a new transport hub as travellers used the new link to shuttle between the five continents.

Gilpin was killed after a collision[④] with a horse and buggy[⑤] in 1894 before he could do anything practical to further the idea of a Bering Strait crossing. But he'd started something, and at about the same time gold was being discovered in Alaska, creating new enthusiasm for plans to improve transport links in the region. In 1892 Joseph Strauss, the man behind San Francisco's Golden Gate Bridge, drew up designs for a bridge to span[⑥] the Strait. But they never got off the drawing board. His critics said a bridge would be too difficult to build and dangerous for traffic in such a cold climate.

注释

① transcend [træn'send] v. 超出，超越（通常的界限）
② intermingle [ɪntə'mɪŋg(ə)l] v. 使（人、思想、色彩等）混合
③ obliterate [ə'blɪtəreɪt] v. 毁掉，覆盖，清除
④ collision [kə'lɪʒ(ə)n] n. 碰撞（或相撞）事故
⑤ buggy ['bʌgɪ] n.（常指无顶无门的）专用小汽车
⑥ span [spæn] v. 持续，贯穿

年，他写了《世界性的铁路：将世界各大陆紧密融合为一体》一书。在这本书中，吉尔平叙述了如何修建一条横跨美洲、亚洲、欧洲的铁路，从而免去横渡大西洋的长途跋涉。这样就可以"超越世界地理的不和谐"，"最终把地球上的所有的人们融合在一起，极大地消除种族差异，使人类成为兄弟姐妹"。

首先要修建一条铁路穿过阿拉斯加（美国于1867年从俄罗斯购得的边境地区）的冰原，直达白令海峡（以1728年穿过这里的丹麦探险家维他斯·白令的名字命名），再在地球的另一侧修建一条类似的铁路，把乘客带到杰日尼奥夫角的俄罗斯海岸，这是亚洲大陆的最东端。乘客然后再乘坐铁路渡船穿过海峡。这样就可以乘火车游历亚洲、欧洲、非洲、南北美洲。作为一位优秀的科罗拉多州人，吉尔平认为，随着旅客不断通过这条新铁路穿梭于五大洲之间，科罗拉多州首府丹佛必将成为新的交通枢纽。

1894年，吉尔平在一场马车事故中不幸身亡，还没来得及为进一步实现横穿白令海峡采取一些实际举措。但他已经开始做一些事了，当时正好在阿拉斯加州发现了金矿，激起了制订计划改善这一地区运输网的新高潮。1892年，金门大桥的设计师约瑟夫·施特劳斯（Joseph Strauss）绘制了横跨白令海峡大桥的设计图，但这些设计图最终未能付诸实施。批评者认为，在这么寒冷的气候下很难架桥，而且交通也很危险。

1902年，英国探险家哈里·德·温特（Harry de Windt）尝试进行了一次从巴黎到纽约的约28967千米（18000英里）的陆上旅行，途中要乘雪橇穿过部分冰封

注释

① husky ['hʌskɪ] n. 因纽特犬
（高大强壮，毛厚，用来拉雪
橇）

② tunnel ['tʌnl] v. 开凿隧道，
挖地道

③ shaft [ʃɑːft] n.（电梯的）升
降机井，通风井，竖井，井筒

In 1902, a British explorer called Harry de Windt attempted to make a 18,000-mile overland journey from Paris to New York via a partly frozen Bering Strait with huskies①, to survey whether a railway link would work. When the weather played havoc with his plans, he'd had to catch a lift in a ship to cross the Strait instead. De Windt called it the worst journey of his life. But on his celebrated arrival in the Big Apple, he said that his experience of the conditions told him only a tunnel, for what he called an All-World Railway, would work. He explained: 'As to the possibilities of the passage for railroad travel I am much more firmly convinced than I hoped to be. I fully believe that the road will be completed and in operation within 12 years. The Bering Strait can be tunnelled②!' De Windt said that the Diomede Islands, which lay half way across the Strait, could be the site of ventilation shafts③ for a rail tunnel. He got as far as an audience with President Theodore Roosevelt about the idea, but no further.

In 1905 a French scientist called Baron Loicq de Lobel decided to approach the Russian ruler Tsar Nicholas II with a $300 million plan for a tunnel and connecting railroad. Acting as a promoter for American businessmen Edward Harriman and James Hill, he imagined: 'No more seasickness, no more dangers of wrecked liners, a fast trip in palace cars with every convenience'. The proposal was to build a tunnel, and a 2,500-mile railroad from it, to link up with the new Trans-Siberian Railway which was being pushed east to west across Russian territory through 5,000 miles of wilderness – an incredible feat for a country still steeped in poverty. For the Tsar, who had just lost a war against the Japanese, the tunnel had the advantage that it might strengthen the country internationally. In 1906 it was reported that he'd given De Lobel's scheme the go ahead.

The idea of crossing the Bering Strait still had many critics

的白令海峡，以考察铁路网是否可行。恶劣天气完全打乱了他的计划，他只好搭船穿过白令海峡。德·温特称这是他一生中最糟糕的一次旅行，但在他到达纽约市举行庆典时，他说这段经历告诉他，只有一条隧道可以穿过白令海峡，成全他所谓的环球铁路。他解释说："关于铁路旅行通道的可行性问题，我不仅仅是希望，而是非常坚信。我完全相信这条铁路将在12年内完工并投入运营。白令海峡可以打隧道。"德·温特说，位于白令海峡中央的狄奥米德岛可以作为铁路隧道的通风井站。他还在会见西奥多·罗斯福（Theodore Roosevelt）总统时谈过这个想法，但没有下文。

1905年，法国科学家卢瓦克·德·洛贝尔男爵（Baron Loicq de Lobel）决定向沙俄统治者沙皇尼古拉二世（Nicholas Ⅱ）提出一项3亿美元的计划，旨在打通隧道和连接铁路。作为美国商人爱德华·哈里曼（Edward Harriman）和詹姆斯·希尔（James Hill）的推销人，他描绘说："不再有晕船，不再有沉船的危险，可以坐着豪华列车方便地旅行。"他的提议是建一条隧道约4000千米（2500英里）长的铁路，与新的西伯利亚铁路相连。西伯利亚铁路穿过约8000千米（5000英里）荒原，横跨俄国，对一个贫穷的国家来说，这无疑是一件了不起的伟绩。对于刚在与日本人的战争中打输了的沙皇来说，这条隧道的优点在于可能会增强俄国的国际影响力。1906年，有报道称沙皇已批准了德·洛贝尔的方案。

in America. The engineering challenge would be significant. As one journalist observed, the tunnel would have to be much longer than the longest tunnel which then existed – the 12-mile Simplon Tunnel in the Alps. The *New York Times* wasn't convinced, saying that, at this point, building a tunnel was: 'about as practical as a plan to colonise the dark side of the moon.' De Lobel needed more than the Tsar's say so to proceed with what he called his Trans-Alaska-Siberian Railway; he needed a formal agreement between the sovereign[1] powers involved.

Despite the apparent agreement of 1906, the next couple of years saw the Tsar and his government blow hot and cold over the strategic merits of the tunnel. The Tsar, it was said, wondered what would happen to the tunnel in a time of war (echoing fears in Britain over the first Channel Tunnel). Not only was the Tsar apparently changing his mind, the money markets in America weren't excited by de Lobel's plan and sufficient funding doesn't appear to have been forthcoming. By the end of the decade, the Bering Strait crossing was lost amid the growing mood of hostility between the great powers.

It was to be some time before anyone realistically considered the idea again. But the outbreak of the Second World War briefly gave it new life. The US government was worried about the possibility of a Japanese invasion through Alaska. So they gave orders for the construction of the Alaskan Highway, a road linking the territory through Canada to the US. It was completed in 1942 and though the highway didn't go as far as the Bering Strait, talk of a tunnel grew again, especially since the Soviet Union was now an ally. This time it was for a subterranean[2] road rather than a railway.

The Cold War was soon to close the door to a link once again. But since the break up of the Soviet Union there has been

注释

① sovereign ['sɒvrɪn] *n.* 君主，元首

② subterranean [ˌsʌbtə'reɪnɪən] *adj.* 地下的

　　跨越白令海峡的想法在美国仍然受到了众多批评。在工程方面面临着严重挑战。一位记者评论说，这条隧道比当时最长的隧道——阿尔卑斯山约19千米（12英里）长的辛普伦隧道长得多。《纽约时报》也不相信，评论建这条隧道的"可行性大概和在月球背面建立殖民地的计划一样"。德·洛贝尔需要的不仅仅是沙皇口头说让他建所谓的"阿拉斯加—西伯利亚大铁路"，他更需要有关主权国家的正式同意。

　　尽管沙皇于1906年明确同意了，但随后几年，沙皇及其政府对这条隧道战略价值的态度忽冷忽热。据说沙皇不知道这条隧道在战时会怎么样（回顾一下英国对第一条英吉利海峡隧道的恐惧）。不久沙皇就变卦了，美国的资本市场对德·洛贝尔的计划也不太感兴趣，于是他没有筹到足够的资金。到了20世纪第一个十年的末期，跨越白令海峡的想法逐渐由于大国之间不断增加的敌对情绪而消失。

　　在之后的很长时间内，一直没有人再次现实地考虑过这一想法，但第二次世界大战再次给了它生机。美国政府担心日本入侵阿拉斯加，于是下令修建阿拉斯加高速公路，通过加拿大把边境与本土连接起来。这条高速公路建成于1942年，由于离白令海峡不远，人们再次提起那条隧道，特别是此时苏联已经成为美国的盟国。这一回，人们考虑的是一条地下公路，而不是地下铁路。

　　"冷战"很快使这一铁路网成为泡沫，但自从苏联解体后，人们又开始讨论通过海峡把两个大陆

注释

① envisage [ɪnˈvɪzɪdʒ] v. 想象，设想，展望
② infrastructure [ˈɪnfrəstrʌktʃə] n.（国家或机构的）基础设施，基础建设

renewed discussion of joining the two continents with a tunnel. Academics have suggested that using it for both transport and oil would be economically beneficial. Even former Russian president Vladimir Putin has expressed his support for such a project. At no more than twice the length of the Channel Tunnel, the technology certainly exists to get it done. And a Bering Strait crossing is certainly more feasible than a tunnel across the 3,000-mile Atlantic which would achieve the same purpose of connecting the Americas with Europe, Asia and Africa. Such an idea was once suggested by Michel Verne, son of the explorer Jules Verne. In 1895 he wrote a piece envisaging① how, one day, steam-driven fans would propel trains through a 3,000-mile Atlantic tunnel at 1,000 mph.

Huge obstacles would lie in the way of such a tunnel. But there's a long way to go before regular traffic starts passing underneath the Bering Strait either. As it stands the infrastructure② doesn't exist to get people to the watery impasse even if there was a tunnel. And long-distance flights have now gone some way to achieving Gilpin's original vision.

相连的想法。学术界认为，把隧道用于交通和石油运输将在经济上十分有利，甚至俄罗斯总统弗拉基米尔·普京（Vladimir Putin）也表示支持这一项目。这条隧道长度不足英吉利海峡隧道的两倍，技术上绝对不成问题。而跨越白令海峡比同样连接美、欧、亚、非各大洲的约4800千米（3000英里）的大西洋隧道更为可行。探险家儒勒·凡尔纳（Jules Verne）的儿子米歇尔·凡尔纳（Michel Verne）曾经提过这一设想。1895年，他写了一篇文章，描述了将来有一天蒸汽动力螺旋桨如何推动火车以约1600千米/时（1000英里/时）的速度通过3000英里长的大西洋隧道。

修建大西洋隧道自然困难重重，但要通过白令海峡地下实现常规交通也同样任重道远。因为即使有了隧道，仍然没有让人们到达这一海峡的基础设施。而长途航班现在已经换了一种方式实现了吉尔平的最初愿景。

~~CANCELLED~~

BRITISH RAIL'S FLYING SAUCER & THE 'GREAT SPACE ELEVATOR'

It would certainly have been difficult to persuade anyone who had the misfortune to consume a British Rail sandwich in the 1970s that the network's service could ever be out of this world. So it comes as something of a surprise that just a year after the first moon landing in 1969, a patent with the number 1310990 was lodged by British Rail for a space vehicle. The man behind this bemusing①, flying-saucer②-shaped craft was scientist Charles Osmond Frederick, who worked at the British Rail research department in Derby. He provided detailed drawings for a 120ft vehicle in which passengers would be ferried into space in a compartment above engines powered by 'a controlled thermonuclear③ fusion reaction ignited④ by laser beams'.

It wasn't an April Fool's joke. When the patent was rediscovered in later years it emerged that British Rail had put its name to the idea simply because it had been done by an employee. They didn't, so they maintained, have any immediate plans for a branch line into space. Scientists who have looked at the blueprint since say the technological worthiness of such a craft is questionable, to say the least. The episode⑤ does,

英国铁路公司的 "飞碟"
和大太空升降机

要想让20世纪70年代不幸消费过英国铁路公司三明治的人相信交通运输网络服务曾经延伸到地球之外，绝对不是易事。在1969年人类首次登陆月球之后的那一年，英国铁路公司为太空列车申请了专利号为1310990的专利。这让人们多少有些惊讶。研制这艘迷人的飞碟状飞行器的人是科学家查尔斯·奥斯蒙德·弗雷德里克（Charles Osmond Frederick）。他在英国铁路公司位于德比的研究部门工作。他绘制了这台长约37米（120英尺）的列车的详细图纸。乘客坐在车舱内，通过下方 "激光束点燃的受控热核聚变反应" 动力发动机被送入太空。

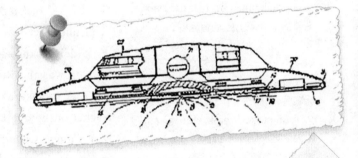

▲ **英国铁路公司于1970年注册专利的飞碟状飞行器。**

A patent for a flying-saucer-style craft, lodged by British Rail in 1970.

这可不是愚人节的玩笑。当人们数年后发现这项专利时，才知道英国铁路公司为这一想法冠名仅仅因为这是其员工的杰作。他们一直没有向太空开通支线

注释

① log [lɒg] n. （某时期事件的）正式记录，日志，（尤指）航海日志，飞行日志

② simultaneously [ˌsɪml'teɪnɪəslɪ] adv. 同时地

③ crater ['kreɪtə] n. 火山口，弹坑

④ orbital ['ɔːbɪt(ə)l] adj. （行星或空间物体）轨道的

⑤ reclusive [rɪ'kluːsɪv] adj. 独处的，隐居的

⑥ scintillating ['sɪntɪˌleɪtɪŋ] adj. （谈话或表演）生动有趣的

⑦ orbit ['ɔːbɪt] n. （天体运行的）轨道

⑧ pension ['pɛnʃən] n. 养老金，抚恤金

⑨ multistage ['mʌltɪˌsteɪdʒ] adj. （火箭或飞弹）多级的

⑩ ignite [ɪg'naɪt] v. 点燃，着火

⑪ extinguish [ɪk'stɪŋgwɪʃ] v. 使破灭，消除

however, prove the enduring romance humans have with attempts to get beyond the Earth's atmosphere.

The first record of anyone actually trying to do so has been credited to a Chinese man called Wan Hu. History logs① him as the world's first astronaut. Well, attempted astronaut. The story goes that in around 1500 he attached forty-seven rockets to a chair and had them lit, simultaneously②, by his servants. Sadly the resulting explosion left no trace of Wan, though he does now have a crater③ named after him on the dark side of the moon. He was, as it turned out, on the right lines. We now know that it was the pioneers of rocket technology that finally enabled man to get into space. The first sub-orbital④ trip was made by a German V-2 rocket during World War Two, and the first human being into space was Yuri Gagarin in 1961 aboard Vostok Ⅰ.

Widely acknowledged as a leading light of early rocket science was the Russian Konstantin Tsiolkovsky, born in 1857. A reclusive⑤ figure, he spent most of his life as a teacher occupying a remote log cabin while working on his scintillating⑥ scientific ideas. Tsiolkovsky never built an actual rocket. But his theories and formulas did lead the way for others to do so in the early decades of the twentieth century. His 1903 work *The Exploration of Cosmic Space by Means of Reaction Devices* explained how a liquid-fuelled rocket could achieve orbit⑦ in space and, late in life, he was given a government pension⑧ to work on his theories, which included the idea of using multistage⑨ rockets to get into space.

By the end of the 1920s his importance in the field was being recognised. Hermann Oberth, a famous German scientist, wrote to Tsiolkovsky saying: 'You have ignited⑩ the flame, and we shall not permit it to be extinguished⑪; we shall make every effort so that the greatest dream of mankind might be fulfilled.' Tsiolkovsky, who died well before the rockets he foresaw

的计划。当时看过设计蓝图的科学家说，至少这种飞行器的技术价值值得怀疑，然而，这一设想确实证明了浪漫的人类始终在尝试飞出大气层。

据史料记载，最早真正尝试飞出大气层的人当属中国人**万户**。史书记载他是世界上第一位航天员，或更准确地说，想成为航天员的人。大约在1500年，他把47级火箭绑在椅子上，然后让人把这些火箭同时点燃。不幸的是，产生的爆炸让万户踪迹全无，不过后来人们以他的名字命名了月球背面的一个陨石坑。最终证明他这种思路是对的。我们现在知道，正是火箭技术的先驱使人类最终进入太空。首次亚轨道飞行是在第二次世界大战期间由德国的V-2火箭完成的，而进入太空的第一人是尤里·加加林（Yuri Gagarin），他于1961年乘载"东方1号"进入太空。

人们一般认为，火箭科学的启蒙者是俄罗斯人康斯坦丁·齐奥尔科夫斯基（Konstantin Tsiolkovsky）。他出生于1857年，是一位十分低调的人，大部分时间都在从事教学工作，同时在一间小木屋里研究他的科学观点。齐奥尔科夫斯基从来没有建造过一艘真正的火箭，但他的理论和方程式却为20世纪之初建造火箭的那些人指明了方向。1903年，他出版了《采用火箭装置探索宇宙空间》一书，书中阐述了如何采用液体燃料动力火箭进入太空轨道。后来，他获得了一笔政府资助，用于继续研究相关理论，其中包括采用多级火箭进入太空的观点。

到了20世纪20年代末，他在这一领域的重要性

注释

万户（？—1390），原名陶广义，后被朱元璋赐名"成道"。原是浙江婺城陶家书院山长，喜好钻研炼丹技巧。一次炼丹事故后，他开始试制火器。万户是第一个想到利用火箭飞天的人，被称为"世界航天第一人"。

actually reached space, even looked ahead to a time when new technology would have to replace them. He predicted that the business of launching rockets would be extremely expensive and thus came up with an idea to tackle the problem: the 'great space elevator'.

The idea still seems far-fetched today, bringing to mind visions of Willy Wonka setting off for the inky^① depths with Charlie Bucket in Roald Dahl's 1972 book *Charlie and the Great Glass Elevator*. But it was the newly built Eiffel Tower and its brand new lifts speeding visitors to the top that convinced Tsiolkovsky that a version to take man into space could be more than a fictional fantasy. The Parisian landmark led him to formulate the idea of a tower, or cable, which would be tethered^② to the planet, extending 22,300 miles upward. In his 1895 work, *Dreams of Earth and Sky*, he spelled out how, at the top, there would be a 'celestial^③ castle' – a kind of space station. This would be held in a geosynchronous orbit – one where the speed of the orbit matches the rotational speeds of the earth. This way it would be far enough up and stable enough so that spacecraft could dock with it. Human spaceflight would become easier, more affordable and so promote the exploration of the cosmos^④.

Of course he knew his tower could never be constructed in his lifetime. It would need to be made out of something stronger than the iron which makes up the Eiffel Tower. Even steel wouldn't do. But Tsiolkovsky's space elevator idea may not have been simply pie in the sky. As Tsiolkovsky pointed out, the current expense of ferrying things and people into space using rockets isn't sustainable and so space scientists are looking at the space elevator idea all over again. They think the technology to build one is not only possible but could become reality in just a few decades. The basic science works and new super-strength

注释

① inky ['ɪŋkɪ] *adj.* 墨黑的
② tether ['tɛðə] *v.* 拴住
③ celestial [sɪ'lɛstɪəl] *adj.* 天上的，天堂的
④ cosmos ['kɒzmɒs] *n.* 宇宙

得到了人们的认可。德国著名科学家赫尔曼·奥伯特（Hermann Oberth）给他写信说："您点燃了火箭，我们不应该让它熄灭，我们应该竭尽全力实现人类的最大梦想。"而齐奥尔科夫斯基恰恰死于他所设想的火箭真正进入太空之前，他甚至已经预测到将来有一天会出现一种可以取代火箭的新技术。他预测到发射火箭的费用极其昂贵，因此想出了一个解决这一问题的办法——大太空升降机。

这个想法至今看起来仍然有些超前，让人想起罗尔德·达尔（Roald Dahl）在《查理和大玻璃升降机》中描述的威利·旺卡和查理·巴基特一起前往漆黑深渊的情景。而正是新建的埃菲尔铁塔和铁塔上全新的搭载游客快速到达塔顶的升降机使齐奥尔科夫斯基相信，用类似装置把人类送入太空绝非科幻小说。这座巴黎的标志性建筑使他形成了连接到地球上的塔或缆绳可以向上延伸约35887千米（22300英里）的想法。在他1895年出版的《地球与天空的梦想》一书中，他描述了如何在最顶端建设一座"天空城堡"——太空站的想法。这种天空城堡位于地球同步轨道上，这样轨道速度就与地球旋转速度一致，航天飞机可以在上面安全着陆，人类太空飞行会变得更加便捷和便宜，从而促进宇宙探索。

当然，齐奥尔科夫斯基知道这种塔在他的有生之年绝不会建成。这种塔需要用比建埃菲尔铁塔所用铁更加坚固的材料制成，即使是钢也不行。但齐奥尔科夫斯基的太空升降机的想法也许绝不是空中楼阁。正如齐奥尔科夫斯基指出的那样，目前采用火箭把人和物体送入太

注释

① nanotube ['nænə,tjʊb] *n.* 碳纳米管，纳米管，纳米电子管
② migrate [maɪ'greɪt] *v.* 迁移
③ humanity [hjuː'mænɪtɪ] *n.* 人类，人性

materials like carbon nanotubes① are being investigated while a leading NASA scientist, David Smitherman, has even said that in fifty years' time the technology should have developed sufficiently to allow construction. He says the idea of a space elevator is no longer science fiction. All that is needed, it seems, for Tsiolkovsky's idea to become a reality is the will to do it.

Tsiolkovsky himself thought that human exploration of space was essential for human life to survive. He said:

'The finer part of mankind will, in all likelihood, never perish – they will migrate② from sun to sun as they go out. And so there is no end to life, to intellect and the perfection of humanity③. Its progress is everlasting.'

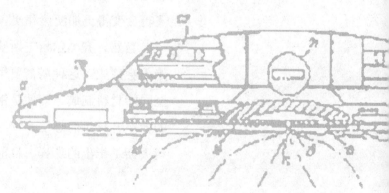

空的成本太高，所以太空科学家开始重新研究太空升降机的想法。他们认为，建设太空升降机的技术不仅有可能性，而且在数十年内有可能成为现实。目前人们正在研究基础科学工程和新型超强度材料（比如碳纳米管），一位**NASA**的主要科学家大卫·斯米泽曼（David Smitherman）曾说过，在50年内应该可以开发出制造太空升降机的技术。他说，太空升降机的想法不再是科学幻想，也许齐奥尔科夫斯基的想法成为现实所需要的只是实现它的意志。

注释

NASA，全称 National Aeronautics and Space Administration，美国国家航空航天局，是美国联邦政府的一个行政性科研机构，负责制定、实施美国的太空计划，并开展航空航天科学研究。

齐奥尔科夫斯基认为，人类探索太空对于人类生存至关重要。他说：

> 人类中的优秀分子十有八九不会消亡——他们会从一颗恒星走向另一颗恒星，所以生命不息，智慧不息，人类的完善不息。这一过程永无止境。

WHY THE BATHYSCAPHE WAS
THE DEPTH OF AMBITION

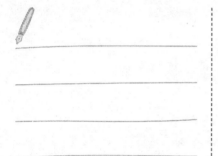

注释

① conquer ['kɒŋkə] v. 征服，攻占
② ignorant ['ɪgnərənt] n. 无知的
③ canyon ['kænjən] n. 峡谷
④ vessel ['vɛsəl] n. 船，舰
⑤ bathyscaphe ['bæθɪskæf] n. 深海潜水器，深海潜水艇
⑥ seabed ['sibɛd] n. 海床

On six occasions, human beings have visited the moon. And the number of people who have spent time in space is measured in hundreds. Yet while man's quest to conquer① other worlds continues, we still remain largely ignorant② of what lies beneath on our own planet. Just two people have been to the deepest reaches of the earth's most inhospitable ocean canyon③, almost 7 miles underwater. Doing so required a hugely expensive, highly effective contraption that reached parts that other vessels④ could not.

Bathyscaphe⑤ (or bathysphere as it is sometimes called) combines two Greek words in a construction that means 'deep boat', but this does the invention little justice. For a bathyscaphe called the *Trieste*, a magnificent deep-water vessel designed by Swiss father and son team Auguste and Jacque Piccard, was to reach the bed of the Mariana Trench in the western Pacific. Any craft risking such a journey, the deepest part of Earth's seabed⑥, must withstand pressure of 17,000lbs. Any structural weakness means certain death to those aboard.

Having already been to the edge of space in a self-

为什么深海潜水器象征雄心·的深度

　　人类6次到达过月球，而到过太空的人数以百计。尽管人类征服其他星球的步伐还在继续，但我们对地球底下的大部分却仍然一无所知。只有2个人曾经到达过地球上最险恶的海沟的最深处，大约位于水下11.3千米（7英里）。只有用一种特别昂贵、十分有效的精巧装置，才能到达其他船只无法到达的深度。

　　"深海潜水器"（又名深海潜水球）一词把两个希腊单词连在一起，意思是"深水船只"，但这个词显然没有准确体现这项发明的价值。因为瑞士父子兵奥古斯特（Auguste）和雅克·皮卡尔（Jacque Piccard）设计的深海潜水器"的里雅斯特号"潜到了西太平洋的马里亚纳海沟深处，这可是一艘巨大的深水船只。要在地球海床的最深处进行这样一次历险，任何舰艇都必须能承受住约1195千克/平方厘米（17000磅/平方英寸）的高压。结构上一旦存在缺陷，就意味着葬身海底。

　　奥古斯特·皮卡尔曾经乘坐自己设计的气球到达过太

注释

① ascend [ə'sɛnd] v. 上（山、楼梯）
② cosmic ['kɒzmɪk] adj. 宇宙的
③ dive [daɪv] n. 潜水
④ torpedo [tɔːˈpiːdəʊ] n. 鱼雷
⑤ squall [skwɔːl] n. 突起的狂风（常引起短暂暴雨或暴雪）
⑥ ballast ['bæləst] n. 压舱物
⑦ pellet ['pɛlɪt] n. 小团，丸
⑧ magnetic [mæg'nɛtɪk] adj. 有磁性的

designed balloon, Auguste Piccard's attempt to go deeper than anyone has done before, or since, was to be his second major adventure. In 1932 he ascended① 16,650m to study cosmic② rays from the stratosphere. Turning to the seas with his son, also an accomplished physicist, and having secured funding from the Italian city of Trieste – hence the name of the vessel – they built their bathyscaphe. And, in progressively more challenging dives③, it worked.

More diving bell than boat, with an appearance combining submarine with torpedo④, the bathyscaphe's first significant journey took it 3,099m to the sea floor off Ponza, Italy, on 1 August 1953. Just one scientific achievement resulted: the Piccard's proved they could get to the seabed and back and survive. Unfortunately, despite its success, this first bathyscaphe was later destroyed in a squall⑤. A second vessel fared better, while the third, the Trieste, was to go down in US naval history. Fifteen metres long with protective walls 13cm thick, and propelled by electric motors, it resurfaced by releasing ballast⑥ pellets⑦ from magnetic⑧ hoppers, and could survive the deepest reaches.

The greatest challenge to deep sea exploration isn't so much the technology as the cost – and the politics. Needing endless amounts of money to keep the project afloat, the Piccards turned to the United States where the Navy enabled them to improve their vessels. By 1960, they were ready to face their toughest test: the world's deepest ocean canyon. At 10,918m, Mariana Trench is 20 per cent deeper than Everest is high, but while Everest has the advantage that on clear days the views are magnificent, it's a different story on the seabed. After twenty minutes peering out of the tiny windows into the darkness, the Piccards had seen all there was to see: one seabed-dwelling flatfish that they were unable to

空的边缘，这次又尝试到达前人未曾到达的深度，开始他的第二次重要探险。1932年，他升到海拔16650米的高度，研究平流层的宇宙射线。这次他又带着儿子转向大海。他的儿子也是一位杰出的物理学家，他们利用意大利的里雅斯特市赞助的资金建成了深海潜水器，所以用城市的名字命名了这艘潜水器。"的里雅斯特号"起航了，不断挑战着新的潜水深度。

　　"的里雅斯特号"的外观就像潜艇与鱼雷的结合体，看起来更像潜水钟，而不像一艘船。1953年8月1日，这艘潜水器在意大利蓬扎岛附近海底3099米的深度开始了首次重要航行。结果只取得了一项科学成果：皮卡尔父子证明了他们能够到达海底，而且能够活着回来。不幸的是，"的里雅斯特号"后来在一次暴风雨中遭到破坏。第二艘"的里雅斯特号"恐怕要更好一些，而第三艘"的里雅斯特号"则载入了美国海军的史册。这艘潜水器长50米，防护壁厚达13厘米，采用电机驱动，通过从磁性漏斗中释放出压舱石浮出水面，还可以在最深的海底生存。

　　深海探险最大的挑战不是技术，而是资金和政

▲ "的里雅斯特号"——一艘到达地球海洋最深处的潜水研究船。

The Trieste, a diving research vessel which visited the deepest part of the Earth's oceans.

注释

① sub [sʌb] *n.* 潜艇

② judicious [dʒuːˈdɪʃəs] *adj.* 明智的

③ hull [hʌl] *n.* 船体

④ tortuous [ˈtɔːtjʊəs] *adj.* 弯弯曲曲的

⑤ mercy [ˈmɜːsɪ] *n.* 仁慈，宽恕

⑥ sceptical [ˈskɛptɪkəl] *adj.* 表示怀疑的

⑦ cripple [ˈkrɪpəl] *v.* 使受伤致残

⑧ wreckage [ˈrɛkɪdʒ] *n.* （飞机、汽车失事或房屋遭损毁后的）残骸

⑨ crammed [kræmd] *adj.* 塞满的

photograph. Proving only that the bathyscaphe was up to the job, they returned, happy but empty handed, to the surface. And that was that. But for this one historic attempt, man has never returned to such depths.

Back on dry land, US military focus was changing. The Navy thought that, by and large, going to the bottom of the sea to see nothing much was a poor use of precious resources that could be better spent on nuclear missiles. Vice Admiral Hyman G. Rickover, known as the 'father of the nuclear navy' for directing the service towards nuclear subs① after the Second World War, put a stop to further *Trieste* dives.

That appeared to be the end for the *Trieste*. But in a judicious②, if tragic and ironic twist, it wasn't long before Rickover was forced to call the vessel back into action. On 10th April 1963 one of the nuclear-powered submarines for which he was responsible, the USS *Thresher*, sank. Somewhere between 400–800m below the surface, the *Thresher's* hull③ had collapsed under the tortuous④ surrounding pressure. 129 crew lost their lives as seawater flooded in and the *Thresher* exploded. If this looked like a serious situation in itself, it was worse than it seemed. A nuclear reactor was out of control, out of reach and at the mercy⑤ of the elements. A shocked Rickover tried to assure a sceptical⑥ world that a radioactive leak was impossible.

By this time the *Trieste*, still a US Navy ship, was lying redundant in San Diego. Its task now: locate the *Thresher* and recover sufficient parts to discover what really crippled⑦ it. The *Trieste* was fitted out with a new, untested, mechanical arm for a recovery mission and delivered on board the USS *Point Defiance*. In June, in an area thought to contain the wreckage⑧ of the *Thresher*, the *Trieste* began the first of a series of dives. Three men crammed⑨ into the vessel for each

治。要保证项目顺利进行，就需要无穷无尽的资金。皮卡尔父子辗转来到美国，在美国海军的资助下，他们得以不断改进潜水器。到了1960年，他们已做好准备应对最严峻的考验——世界上最深的海沟。马里亚纳海沟深10918米，比珠穆朗玛峰的高度还深20%，而珠穆朗玛峰的优点是在晴天时风景十分壮丽，但海底则不同。皮卡尔父子从小窗里盯着黑暗的海底看了20多分钟，看到的一切就是海底生比目鱼，但他们没法拍摄下来。他们高兴地浮出水面，除了证明深海潜水器能够胜任这项任务，他们简直就是空手而归。事情就是这样，但除了这一次历史性的尝试，人类就再也没有到达过这样的深度。

他们返回陆地之后，美国海军的注意力发生了变化。美国海军认为，到海底去看一些虚无之物，简直是对宝贵资源的浪费，还不如造核导弹。海军中将海曼·乔治·里科弗（Hyman G.Rickover，因其在第二次世界大战后指导海军向核潜艇转变而被称为"核潜艇之父"）叫停了"的里雅斯特号"的进一步潜水活动。

"的里雅斯特号"似乎已经走上穷途末路了。具有讽刺意味的是，里科弗不久之后又把"的里雅斯特号"调回来执行任务。1963年4月10日，他负责的一艘核动力潜艇"长尾鲨号"沉没了。"长尾鲨号"舰体在水下400～800米的巨大水压下破裂，舱内进水，潜水艇发生爆炸，129名船员全部丧生。事态看起来已经够严重了，但更严重的情况是一台核反应堆处于失控状态，完全听天由命。里科弗震惊了，尝试向忧心忡忡的世人保证不会发生核泄漏。

当时"的里雅斯特号"还是美国海军的舰艇，正悠

注释

① encase [ɪn'keɪs] v. 包，围
② pinger ['pɪŋə] n. 声脉冲发送器
③ crew [kru:] n. 全体船员，全体机务人员
④ jiggle ['dʒɪgəl] v. 摇动
⑤ gyrocompass ['dʒaɪrəʊˌkʌmpəs] n. 回转罗盘
⑥ malfunction [mæl'fʌŋkʃən] v. 出故障
⑦ navigation [ˌnævɪ'geɪʃən] n. 航行
⑧ watt [wɒt] n. 瓦特（电的功率单位）
⑨ peer [pɪə] v. 费力地看，盯着
⑩ twist [twɪst] v. 扭曲，拧
⑪ pipe [paɪp] n. 管子
⑫ debris ['deɪbrɪ] n. 碎片，散乱的垃圾
⑬ haul [hɔ:l] v.（用力地）拉
⑭ subsequent ['sʌbsɪkwənt] adj. 随后的
⑮ submersible [səb'mɜ:səbəl] adj. 能在水下操作的
⑯ syntactic [sɪn'tæktɪk] adj. 句法的

trip to the ocean floor. It wasn't easy, and a bathyscaphe isn't a comfortable place to be at the best of times. When you're encased① in a hard metal shell with the possibility of landing on a nuclear reactor or, no less frightening, a nuclear missile, nerves are tested all the more.

Locating the wreckage took several attempts. On the first, small sound generators called 'pingers②' that were meant to guide the *Trieste* into place, failed, directing the crew③ to the wrong area. On the second attempt, the pilot steered the bathyscaphe into the mud on the seabed, where it became firmly stuck. After jiggling④ about for half an hour, the vessel scraped itself loose and hurtled away. On the third attempt, the crew found a plastic shoe cover used by mariners in the reactor compartment of a nuclear sub. At last, they were in the right place, although a fourth dive located nothing. On the fifth attempt, the gyrocompass⑤ broke, the starboard propeller malfunctioned⑥ and the navigation⑦ system closed down, leading to a tow back to Boston for repairs ahead of any further attempt. This wasn't looking good for a cutting-edge machine that had been given a second opportunity to show what it was made of.

Then, on 29 August, success came at last. Although its 4,500 watt⑧ light could peer⑨ only 50ft into the blackness, the bathyscaphe's underwater camera located twisted⑩ metal, including a piece of pipe⑪ marked with the *Thresher's* name. Using the *Trieste's* new mechanical hand, the debris⑫ was hauled⑬ in.

Since those heady days of the 1960s, bathyscaphes have seen little action. Subsequent⑭ ones made by the Chinese reach only a third of the depths achieved by the *Trieste*. The US Navy has a submersible⑮ device called *Alvin*, made in part out of syntactic⑯, buoyancy foam made by General

闲地停泊在圣地亚哥港。它接到的任务是：确定"长尾鲨号"的位置，并回收足够的残骸以调查它的受损原因。

"的里雅斯特号"上配备了新式未经测试的机械臂，用于执行回收任务，后来这个机械臂被装载到美国军舰"迪非恩斯角号"上。1963年6月，"的里雅斯特号"在认为可能散落有"长尾鲨号"残骸的地区开始了一系列下潜中的第一次。三个人挤进狭窄的舰舱，开始在海底航行。真不容易啊！即使在最好的情况下，深海潜水器也不是个舒服地方。当你钻进坚硬的金属壳里，面对着碰到核反应堆或核导弹的可能性时，你的神经时刻经受着挑战。

定位残骸进行了多次尝试。第一次尝试时，采用一台名为"声响器"的声波发生器引导"的里雅斯特号"进入位置，但失败了，导致船员进入了错误的地区。第二次尝试时，领航员把"的里雅斯特号"引到了海底的淤泥里。挣扎了半个小时之后，潜水器终于摆脱了淤泥，狼狈返航。第三次尝试时，船员发现了潜水员在核潜艇反应堆室内所用的鞋套。最后，他们到了正确的位置，但这第四次还是空手而归。第五次尝试时，回转罗盘坏了，右舷螺旋桨出现故障，导航系统也出了问题，只好在进行后续尝试前返回波士顿维修。对于一台获得第二次展现机会的尖端机器来说，这可不太好。

1963年8月29日，定位终于成功了。虽然"的里雅斯特号"4500瓦的灯只能在漆黑的海底照射约15米（50英尺），但潜水器上的水下摄像机还是定位到了扭曲的金属件，包括上面印有"长尾鲨号"名称的一根钢管。新型机械臂最终把残骸拉了出来。

Foods in the same factory where they produce breakfast cereals. But it can only reach depths of 4,500m, a snip[1] of that achieved by the *Trieste*. Despite notable ups and downs, however, it has been infinitely more successful. It located a lost hydrogen bomb off the coast of Spain in 1966 and was attacked by swordfish[2] a year later, forcing an emergency ascent[3] to the surface where the fish was removed from the skin of the vessel and cooked for dinner. The *Alvin* sank the following year but was eventually recovered from the seabed and refitted, going on to discover vast colonies of 3m-long tube worms, a species previously unknown. Although coming up to fifty years old, it remains the planet's most sophisticated[4] aqua-research vessel.

Nuclear submarines still steal through the watery darkness and occasionally get into terminal difficulty, despite the tragedy of the *Thresher*. Today, however, tourism and warfare pay more attention than science to what goes on in the very reaches of the earth. The world has its first tourist submarine, the *Auguste Piccard*, built by Jacques Piccard in honour of his father, and a larger one, the PX-44, for sizable holiday parties. But ultimately, anyone who wants to intimately understand the topography[5] of the ocean floor will be disappointed. Man has better maps of Mars and the moon than the seabed. And there are no plans to change that.

注释

① snip [snɪp] v. 快速剪
② swordfish ['sɔːdˌfɪʃ] n. 箭鱼
③ ascent [ə'sɛnt] n. 攀登
④ sophisticated [sə'fɪstɪˌkeɪtɪd] adj. 高级的，复杂的
⑤ topography [tə'pɒɡrəfɪ] n. 地形学

20世纪60年代那些令人兴奋的日子之后，深海潜水器就没什么作为了。**后来中国人制造的深海潜水器曾下潜到了"的里雅斯特号"下潜深度的1/3处。**美国海军还有一种名为"埃尔文号"的潜水装置，部分零件是由通用食品公司在其生产早餐谷物的工厂里采用合成浮力泡沫制成的。但它只能下潜4500米，与"的里雅斯特号"相比完全不是一个量级的，不过它显然更为成功——它曾于1966年在西班牙海岸边定位了一颗遗失的氢弹。一年后，"埃尔文号"受到一条箭鱼的袭击，被迫紧急浮出水面，把鱼从船体上取下来做了午餐。第二年"埃尔文号"沉没了，但后来又被从海底捞出来修好。接着它发现了一种长3米的管状蠕虫组成的群落，这种蠕虫是一种以前不为人知的物种。在"埃尔文号"50岁高龄时，它仍然是地球上最先进的水下研究船。

尽管出现了"长尾鲨号"的惨剧，但核潜艇仍然偷偷地潜伏在黑暗的水下，偶尔遇到终结其生命的困难。时至今日，旅游界和军界都比科学界花在地球深处的精力要多。世界上有了第一艘观光潜艇"奥古斯特·皮卡尔号"，这是雅克·皮卡尔为了纪念自己的父亲而建成的。而更大的一艘观光潜艇PX-44号则适合规模较大的假日旅行团。不过直到最后，渴望清晰了解海底地形的人还是会失望——绘制的火星地图和月球地图都比海底地图要清楚，而且人类目前尚无计划对此做出改变。

编者注
2012年6月27日，中国的"蛟龙号"载人潜水器曾到达水下7062.68米的深度。

215

DISASTER

BRUNEL'S NOT SO *GREAT EASTERN*

With two record-breaking steamships behind him, engineer, entrepreneur and showman Isambard Kingdom Brunel needed an idea bigger, better and bolder[①] to continue pushing the boundaries of engineering. But the result, his next ship the *Leviathan*, later known as the SS *Great Eastern*, proved ruinous[②]. Arguably, it killed him.

A monster of a ship conceived in 1852, Brunel's *Leviathan* would demolish[③] records by the score. At 211m long and weighing[④] in at 22,500 tons, its volume would be six times larger than any existing vessel. It would displace more water and sport more masts, sails, furnaces and crew. In a fair wind, it would steam its way non-stop across 22,000 miles of ocean; almost enough to get the whole way round the globe without refuelling[⑤]. And by travelling great distances faster and providing comfortable, air-conditioned cabins deep within the hull, it would also be a commercial success. 'I never embarked[⑥] on one thing to which I have so entirely devoted myself, and to which I have devoted so much time, thought and labour, on the success of which I have staked[⑦] my reputation', Brunel wrote later. This was prescient[⑧]; for by the time the ship sailed, his

布鲁内尔名不副实的
"大东方号"

一场灾难

爱出风头的工程师、企业家伊萨姆巴德·金德姆·布鲁内尔（Isambard Kingdom Brunel）已经有了两艘打破纪录的蒸汽船，他还需要一个更妙、更好、更大胆的主意，继续推进工程的界限。他设计的另一艘船"利维坦号"，后来称为"大东方号"汽船，最后却成为毁灭性的存在，而且可以说是害死了他。

布鲁内尔的"利维坦号"构思于1852年，这个庞然大物打破了多项纪录。它长211米、重22500吨，体积比当时最大的船大6倍。它的排水量更大，配备有更多的桅杆、帆、锅炉和船员。在顺风的情况下，它能一口气航行约35400千米（22000英里），几乎不补充燃料就能绕地球一圈。船体内安装了空调，即使在远距离高速航行的情况下也很舒服，它在商业上也非常成功。后来布鲁内尔写道："我从未在一件事上倾注过如此多的心血，从未倾注过如此多的时间、思考和劳动，我用自己的名声担保它的成功。"他真有预见性，因为船出海

注释

① tatter ['tætə] v. 撕碎

② curse [kɜːs] v. 诅咒

③ prospectus [prə'spɛktəs] n. (公司或学校的)介绍说明文件

④ bait [beɪt] n. 诱饵

⑤ dockyard ['dɒkˌjɑːd] n. 造船厂，修船所

⑥ cynically ['sɪnɪklɪ] adv. 牟取私利地

⑦ bid [bɪd] n. 出价，投标；努力争取

⑧ ludicrous ['luːdɪkrəs] adj. 荒谬的

⑨ magnitude ['mægnɪˌtjuːd] n. (尺寸、规模、重要性等)大的程度

⑩ forewarn [fɔːˈwɔːn] v. 预先告知(即将发生之事)

⑪ caution ['kɔːʃən] v. 警告，告诫

⑫ dock [dɒk] n. 码头

⑬ mast [mɑːst] n. 桅杆

⑭ quip [kwɪp] n. 俏皮话，妙语

⑮ woeful ['wəʊfəl] adj. 悲伤的

⑯ spiral ['spaɪərəl] v. 使……螺旋式生长或移动，螺旋式生长或移动

⑰ mount [maʊnt] v. 组织，发动

reputation was in tatters①.

From start to finish, the *Leviathan* appeared cursed②. Firstly, the Eastern Steam Navigation Company, which was talked into ordering the ship from the ever-persuasive Brunel, was left with nothing to use it for when it lost a government contract (Brunel's prospectus③ was nothing more than 'gross bait④', rued the chief constructor of Woolwich dockyard⑤). Then shipbuilder John Scott Russell, delighted when the famous engineer accepted his remarkably, possibly cynically⑥, low bid⑦ for construction, began to cut corners to meet the ludicrously⑧ tight budget. But most significantly, the choice of shipyard proved disastrous for a vessel of such magnitude⑨.

Early in the project, Brunel had been forewarned⑩ of launch difficulties by the Institute of Civil Engineers. The ship could never launch safely in the Thames at the Isle of Dogs, they cautioned⑪ – the width of the river at this point being not much greater than the length of the ship. But securing dock⑫ facilities with a launch site sufficient to handle the world's heaviest ship would add to costs. Ignoring the civil engineers' advice, Brunel built the *Leviathan* at a dock from where it would be unable to launch safely. Nonetheless, as the shell began to take shape, its magnificence could not be doubted. The double hull – with 30,000 iron plates, each uniquely shaped, stamped with a number and then fitted together in the fashion of a giant jigsaw puzzle – was the kind of engineering for which Brunel was famed. The six giant masts⑬ were each given the name of a day of a week, from Monday to Saturday, allowing the crew to quip⑭ that they would never have a Sunday at sea.

But all was not well. Scott Russell's undoubted skills as a shipbuilder proved insufficient to compensate for his woeful⑮ underestimation of the complexities of the *Leviathan's* construction. As costs spiralled⑯, his debts began to mount⑰,

时，他也随之名声扫地了。

"利维坦号"自始至终看起来都有点倒霉。首先，当东方航运公司从能言善辩的布鲁内尔订购它的时候，这家公司由于丢掉了一个政府合同，其实已经一无所有了。布鲁内尔的计划书无非是个诱饵——伍尔维奇船坞的总工程师懊悔地表示。其次，当这位著名的工程师接受了造船工程师约翰·斯科特·拉塞尔（John Scott Russell）明显过低的工程报价后，拉塞尔非常高兴，就开始偷工减料，从而弥补紧张得可笑的预算。最重要的是，船坞的选择对于这么大的船来说简直就是一场灾难。

项目早期，英国土木工程师学会就向布鲁内尔警告了下水的难题。他们警告说，这艘船绝不能在泰晤士河的爱犬岛处安全下水，泰晤士河在这里的宽度还没有船的长度长。但要保证船坞设施具备世界上最重的船的下水点会增加成本。于是布鲁内尔无视土木工程师学会的建议，在一个无法安全下水的船坞建起了"利维坦号"。纵然如此，随着船壳开始成形，船的宏大实在毋庸置疑。这艘船是双层船壳，共用了3万块铁板，每块铁板形状不同，上面印有数字，然后用拼巨型七巧板的方式组装在一起，这正是布鲁内尔为之自豪的那种工程。6根巨大的桅杆分别以星期一到星期六命名，因此船员讽刺说他们在海上肯定过不上星期天。

然而事情并非一帆风顺。斯科特·拉塞尔不容置疑的造船能力也无法弥补他对"利维坦号"施工复杂性的可悲低估。随着成本螺旋式上升，他开始债台高筑，后来供货商都不愿意与他共事了。布鲁内尔也不帮忙，根据历史学

until eventually suppliers were unwilling to work with him. Brunel didn't help, 'throwing his weight around' and becoming 'critical, petulant① and downright② rude in his dealings with him' according to historian George Emmerson. Because Brunel only paid Russell on the weight of iron erected, and a quarter of that payment was in worthless shares, bankruptcy loomed. At one point, with Russell's creditors within their rights to seize the ship and Brunel believing he had extracted all the value he could out of him, Brunel chivvied③ the Eastern Steam Navigation Company into accepting early delivery of the ship. A quarter of the hull remained incomplete, fitting out was some way away, and there was still the problem of that looming launch in the Thames. If that wasn't bad enough, disaster was to follow.

Tragedy had already visited the *Leviathan*. A work boy, falling head first onto a spike, died horribly. And working away in a watertight and, as it turns out, unfortunately airtight cell between the gaps of the double hull, one riveter④ slowly suffocated⑤, his yells going unheard as a thousand colleagues banged⑥ away in adjacent⑦ cells. His death, it is said, brought the bad luck to the ship, as one visitor discovered when his own head was crushed in as he looked around.

But industrial accidents, even fatal⑧ ones, did not derail construction unduly⑨ in Victorian England, and before long Brunel decided the time had come to launch. To overcome the problems of the site, this would be done sideways and at an angle that would lift the ship's bow 12m. Taking advantage of a high tide on 7 November 1857, and with the ritual⑩ crash of champagne against its side, the *Leviathan* launched. Anticipating problems, Brunel was eager to keep crowds away. But you can't launch the biggest ship the world has ever seen, on which thousands of workers have been employed, and which has generated such publicity (not all of it good) and hope that

注释

① petulant ['pɛtjʊlənt] adj. 使小性子的

② downright ['daʊnˌraɪt] adv. (强调不快或负面事物) 彻头彻尾地

③ chivy ['tʃɪvɪ] v. 使烦恼

④ riveter ['rɪvɪtə] n. 铆钉枪，铆工

⑤ suffocate ['sʌfəˌkeɪt] v. 使窒息而死，窒息而死

⑥ bang [bæŋ] v. 砰然作响

⑦ adjacent [ə'dʒeɪsənt] adj. 相邻的

⑧ fatal ['feɪtəl] adj. 后果严重的

⑨ unduly [ʌn'djuːlɪ] adv. 过分地

⑩ ritual ['rɪtjʊəl] adj. 仪式性的，传统的

家乔治·爱默生（George Emmerson）所说，他"开始耍威风"，变得"非常苛刻，性情暴躁，在与他共事时十分粗暴"。由于布鲁内尔只根据所安装铁块的重量给拉塞尔付款，而且1/4的付款属于毫无价值的股权，于是破产迫在眉睫。就在这时，拉塞尔的债权人按照权利查封了这艘船，而布鲁内尔认为他已经倾囊而出，于是就不停骚扰东方航运公司，要求接受提前交船。这时，1/4的船壳尚未完工，装配工作正在进行，而且在泰晤士河下水还隐约有些问题。然而这还不够糟糕，灾难即将来临。

其实悲剧已经降临"利维坦号"了。一位童工一头摔到一枚长钉上，悲惨地死去。一位铆工在双层船壳缝隙内不透水（不幸的是还不透气）的小舱里连续工作，慢慢地发生窒息，由于周围小舱里的千名同事不停地敲打船体，没人听到他的叫声。据说他的死给这艘船带来了噩运，因为一位访客在四处浏览时头被碾得粉碎。

不过工伤事故，哪怕是死了人的事故也不会叫停维多利亚时期英国的工程建设，而且不久之后布鲁内尔就认为下水的时机成熟了。为了克服场地问题，他决定横向下水，把船头抬高12米，呈一定角度。1857年11月7日，通过充分利用这天的涨潮，伴随着周围庆祝时激情四射的香槟，"利维坦号"下水了。布鲁内尔担心出现问题，希望人群远离。但没有人，就无法使世界上未曾有过的最大的船下水。造船雇用了上千工人，而且产生了那么大的影响力，希望没人注意

注释

① recoup [rɪ'kuːp] v. 弥补，收回

② calamitous [kə'læmɪtəs] adj. 灾难的

③ hydraulic [haɪ'drɒlɪk] adj. 液压的

④ snap [snæp] v. 使咔嚓折断

⑤ lurch [lɜːtʃ] v.（尤指向前）打趔趄

⑥ precarious [prɪ'kɛərɪəs] adj.（情况）不稳定的

⑦ gallon ['gælən] n. 加仑（液体计量单位，1加仑约3.785升）

⑧ cradle ['kreɪdəl] n. 摇篮

⑨ manoeuvre [mə'nuːvə] n. 操控手段

⑩ groan [grəʊn] v. 呻吟

⑪ flee [fliː] v. 逃离

⑫ condemn [kən'dɛm] v. 谴责，责备

⑬ protractor [prə'træktə] n. 量角器，分度规

⑭ hydraulics [haɪ'drɒlɪks] n. 水力学

nobody notices. The enterprising Eastern Steam Navigation Company, anxious to recoup① some of its investment quickly, didn't help by selling 3,000 tickets.

The launch was calamitous②. As restraining chocks were knocked away and hydraulic③ presses began the process of getting the ship into the water, its impressive but inadequate chains snapped④, leaving the *Leviathan* lurching⑤ precariously⑥ on the launch pad. Although Brunel had loaded gallons⑦ of water to the front of the ship to redress an imbalance on the two cradles⑧ of the pad, this only added several hundred tons more weight and caused havoc with the manoeuvre⑨. As it creaked and groaned⑩, it looked like the ship would topple. Workers fled⑪ in terror. Spectators screamed as the hulk slid towards the water. But then, after just a few metres it shuddered to a halt. The damage was done. In the space of just a few minutes, two people had died and several were seriously injured. The sideways launch – and Brunel himself – was condemned⑫ by *Mechanics Magazine*. 'An unnecessary display of self-confidence', it admonished.

Two months later, Brunel tried again, but accepted that the Isle of Dogs location had been a bad idea all round. Even for the famous engineer, choosing another way to launch the *Leviathan* was a puzzle too far. Fellow inventor, the even more influential Robert Stephenson, by this time preoccupied by the process of dying, rose from his bed, got into his dressing gown and slippers and reached for his protractor⑬. More powerful hydraulics⑭ would be needed if this ship was ever to set sail, he said. Yielding 3,500 tons of force in addition to the 100 tons of gravity, the extra hydraulic power worked. On the second attempt on 31 January 1858, the *Leviathan* shamefully slipped into the Thames, tearing up the bed of the river as it did so, with no sign of the large crowds that had risked their lives before.

到实在太难了。东方航运公司急于收回投资，迫不及待地卖了3000张票。

下水造成了灾难。随着固定楔子被撞开，水压机开始把船往水里推，它惊人但力量不足的链子噼啪作响，让"利维坦号"晃晃悠悠地倾斜在下水台上。虽然布鲁内尔在船的前部加了数加仑水，以矫正下水台两个支船架之间的不平衡，但这唯一的作用只是增加了几百吨重量，造成了移动时的浩劫。随着吱吱的响声，船眼看就要倒了。工人们害怕极了，开始逃跑。当笨重的船倾向水中滑动时，观众们尖叫起来。但船仅仅颤抖了几米后，就突然停了下来。惨剧已经发生了。仅仅才几分钟，就死了两个人，多人身受重伤。船只的横向下水和布鲁内尔本人都受到了《机械杂志》的谴责，这本杂志责备说，这是"自信心的不必要展示"。

两个月之后，布鲁内尔又试了一次，但承认爱犬岛这个地点确实不太好。即使对于最著名的工程师来说，要想选择另一种让"利维坦号"下水的方式也太难了。影响力更大的发明家同行罗伯特·史蒂文森当时已经病得奄奄一息，也从病床上爬了起来，穿着睡衣和拖鞋去找量角器。他说，如果船要起航，还需要功率更大的水压机。除了100吨重力外，又产生了3500吨外力，这样增加的水力开始发挥作用。1858年1月31日第二次尝试时，"利维坦号"不光彩地一头栽进泰晤士河，把河床都压坏了，这次人们吸取教训，再也不敢冒险围观了。到这时这艘船已经花了

注释

① maiden ['meɪdən] *adj.*（航行、飞行）首次的

② smash [smæʃ] *v.* 打碎，破碎

③ toss [tɒs] *v.* 扔

④ limp [lɪmp] *v.* 跛行

⑤ anchor ['æŋkə] *v.* 抛锚

⑥ shipshape ['ʃɪp ʃeɪp] *adj.* 整齐而状况良好的

⑦ capsize [kæp'saɪz] *v.* 使（船）倾覆，（船）倾覆

⑧ drown [draʊn] *v.* 溺死

⑨ vicinity [vɪ'sɪnɪtɪ] *n.*（在……）附近

⑩ gale [geɪl] *n.* 大风

⑪ redemption [rɪ'dɛmpʃən] *n.* 救赎，偿还

⑫ drift [drɪft] *v.* 漂流

£1 million had been spent by this time, including £170,000 on the launch, and the ship had still to be kitted out. The Eastern Steam Navigation Company was almost broke, but at least now it could begin to make money.

On 9 September 1859 on its way to Weymouth, from where it would begin its maiden① voyage to New York with paying passengers, disaster struck again. When the ship reached open waters an explosion tore through it, bringing a funnel smashing② onto the deck, shattering cut glass mirrors and tossing③ furnishings around. In the boiler room, five crew were killed. The ship limped④ to Weymouth, where it anchored⑤ for repairs and for the bodies to be removed. The news rocked Brunel. Already ill, he took it badly. Not long after, he was dead, beating his friend Robert Stephenson to the grave by four weeks. He was 53.

It was nearly a year before the *Leviathan*, now called the SS *Great Eastern*, was shipshape⑥ again, but its reputation went before it. Just thirty-five passengers paid for the first voyage on 14 June 1860, and were looked after by 418 crew. To the dismay of the passengers who by that time had boarded, the captain put back the sailing by three days because the crew was drunk. The ship's curse continued to strike. The captain who had taken her into Weymouth after the explosion, William Harrison, later capsized⑦ in one of the ship's boats in a squall near Southampton and drowned⑧. On its second voyage the following year, the boat taking passengers to the ship off Milford Haven ran aground and they had to be rescued by other boats in the vicinity⑨. On the third trip, a gale⑩ rolled the ship, smashing the starboard paddle wheel beyond redemption⑪, causing the port paddle to be lost completely and damaging the rudder. Passengers were injured. For three days the ship drifted⑫ at sea, out of control. It reached the south coast of Ireland,

100万英镑,光下水就花了17万英镑,而船还需要配备其他东西。东方航运公司眼看就要破产了,不过至少现在它可以开始赚钱了。

1859年9月9日,"利维坦号"正驶向韦茅斯,从那里它将载着付费乘客驶向纽约,开始它的处女航。灾难再次降临:当船到达开阔水面时,发生了一起爆炸,把烟囱炸到了甲板上,把刻花琉璃震得粉碎,把设备扔得四处都是。在锅炉房里,还炸死了5名船员。"利维坦号"航行到韦茅斯,停泊后进行维修和尸体清运。这个消息震惊了布鲁内尔,这时他已经病了,感到非常伤心。不久之后,他就去世了,比他的朋友罗伯特·史蒂文森早死了四个星期,时年53岁。

差不多一年之后,"利维坦号"(现在叫"大东方号")又恢复了原样,但它的名声已经臭了。1860年6月14日,它的第一次航行只有35名乘客,而光船员就有418名。让乘客丧气的是,船长因为船员醉酒而推迟起航三天。这艘船的厄运再次降临。曾在爆炸后把船带到韦茅斯的船长威廉·哈里森(William Harrison),后来又在南安普敦附近的一次暴风雨中因乘坐船载小艇而翻船,被淹死了。在第二年的第二次航行中,把乘客从米尔福德港送到"利维坦号"的那艘船又搁浅了,只好让附近的船员把乘客救了出来。在第三次航行中,一阵大风袭击了"利维坦号",把右舷明轮吹得粉碎,根本无法修复,造成左舷明轮完全受损,弄坏了方向舵,还有乘客受伤。于是船失去控制,在海上漂了3天。后来"利维坦号"

注释

① convert [kən'vɜːt] v. 转变
② transatlantic [ˌtrænzət'læntɪk] adj. 横跨大西洋的
③ scrap [skræp] v. 毁

where the harbour master at Queenstown refused the ship entry because the captain was unable to control it. Once again, passengers had to take to small boats to get ashore.

It was the end for the SS *Great Eastern* as a passenger ship, but there was some modest success ahead. After being converted① into a cable-layer, it installed the first transatlantic② telegraph cable in 1866, before becoming a floating music hall. Brunel's *Great Babe*, as he had called her, was finally broken up for scrap③ in 1889 – although being so big, this took eighteen months. *Mechanics Magazine* summed up the venture: 'No failure of his ever did so much to lower the reputation of English engineers as the launch of the *Leviathan*.'

到达爱尔兰的南岸，但皇后镇的港口主任因为船长无法控制船，所以不让船进港。乘客再次乘小船上岸。

这下"大东方号"作为客轮的使命结束了，但前方还有成功在等着它。在成为一座浮动音乐厅之前，"大东方号"转为布缆船，在1866年铺设了第一条跨大西洋电报电缆。布鲁内尔的"大宝贝"（他是这么叫它的），最终于1889年被拆为一堆废铁。由于船太大，光拆就用了18个月。《机械杂志》概括说："他的任何失败也都比不上'利维坦号'下水更给英国工程师丢脸了。"

DISASTER

BESSEMER'S
ANTI-SEASICKNESS SHIP

'Few persons have suffered more severely than I have from sea-sickness.' So wrote Sir Henry Bessemer, the renowned inventor and industrialist, in his autobiography①. On a journey across the English Channel from Calais to Dover in 1868 he'd been so ill that, on his arrival back at his London home, his doctor had attended to him throughout the night. According to Bessemer the doctor had even administered a small amount of prussic acid (actually a poison), to try and cure the malady.

Some seven years after that awful journey Bessemer was at sea once again, this time travelling from Dover to Calais on board a ship named after him – the SS *Bessemer*. The ship had been built with a very specific purpose. It was designed to solve, once and for all, the common ailment② of seasickness which had long afflicted③ millions. Even that hero of the Battle of Trafalgar, Lord Nelson, had been a sufferer.

Sir Henry Bessemer is best known for coming up with a way of making cheap steel. His Bessemer Converter was one of the great machines which helped power the industrial revolution and, later, made skyscrapers possible. By the time his huge paddle steamer the SS *Bessemer* put to sea, the Hertfordshire-

贝西默的防晕船轮船

"很少有人比我晕船更严重的了。"著名发明家、工业家亨利·贝西默（Henry Bessemer）爵士在自传中这样写道。1868年，他有一次穿越英吉利海峡从加来去多佛，中途病得如此厉害，以致回到他在伦敦的家里之后，医生忙了一个通宵来照料他。据贝西默说，医生甚至给他开了少量氢氰酸（一种毒药）用于治病。

那次糟糕的旅行之后七年左右，贝西默又出海了，这次是乘一艘以他名字命名的轮船"贝西默号"从多佛去加来。这艘船建造时就有一个特殊目的，用于彻底解决长期困扰数百万人的常见晕船病。甚至特拉法加海战的英雄纳尔逊勋爵（Lord Nelson）也是晕船患者。

亨利·贝西默爵士最负盛名的是想出了一种廉价的炼钢方法。他的贝西默转炉是推动工业革命的伟大机器，后来又使摩天大楼成为可能。当他制造的这艘

born genius had become internationally famous and very rich. In his lifetime Bessemer had come up with scores of patents covering everything from pencils to artillery① shells. Many of his ideas were unassailable triumphs. His ship that would leave its passengers free from seasickness would prove to be a rather more brittle② proposition③.

It was Bessemer's success that took him to sea, as he travelled abroad on business. But he loathed④ it. Following that dreadful bout of seasickness in 1868 he decided to do something about the problem which he knew affected so many. His brainwave was to fashion a 'swinging saloon' inside a ship that would be set on gimbals⑤, or pivoted supports. Weighted underneath it would move independently of the ship's hull, kept horizontal with the help of hydraulics. This way, he believed, the saloon would be free from the rolling motion of the ship, which was the chief cause of the sickness which afflicted him and others.

Bessemer started work on a large model in his own garden at Denmark Hill in South London. The location was chosen partly because Bessemer couldn't bring himself to test his invention anywhere near the sea. Eventually his model saloon was big enough for people to sit inside. This way he could demonstrate the idea to his peers. Using a steam engine, Bessemer replicated⑥ the movements of a ship at sea and, by all accounts, was able to make his saloon stay level. The inventor reported that his fellow engineers had proclaimed⑦ his experiments a complete success.

A firm, the Bessemer Saloon Steamboat Company, was soon set up to oversee the building of a cross-Channel steamer that would be specially adapted to accommodate his contraption. At 350ft long, the ship would be much bigger than the existing steamers. This length, it was felt, would reduce pitching⑧ at the

注释

① artillery [ɑːˈtɪlərɪ] n. 大炮
② brittle [ˈbrɪtəl] adj. 硬脆易碎的
③ proposition [ˌprɒpəˈzɪʃən] n. 主张，观点
④ loathe [ləʊð] v. 厌恶
⑤ gimbals [ˈdʒɪmbəlz] n. 平衡环
⑥ replicate [ˈrɛplɪkeɪt] v. 复制，重做（试验、工作或研究）
⑦ proclaim [prəˈkleɪm] v. 宣布
⑧ pitch [pɪtʃ] v. 投掷

大型轮船"贝西默号"出海时，这位赫特福德郡出生的天才已经闻名遐迩、腰缠万贯。贝西默一生中发明了许多专利，小到铅笔，大到炮弹。他的许多想法都无懈可击，但这艘旨在让乘客不晕船的轮船却成了一项脆弱的提议。

正是这些成功让贝西默屡屡出海，他要出国做生意，但他十分不情愿出海。在1868年那次可怕的晕船之后，他决定做点事情解决这个困扰许多人的问题。他灵机一动，想起在船内设置一间摇摆交谊厅，安装在平衡环或转动支撑上。下部加重物后，摇摆交谊厅就可以独立于船壳移动，借助液压保持水平。他认为通过这种方法，摇摆交谊厅就可以不受船体颠簸的影响，而船体颠簸正是造成晕船的主要原因。

贝西默在伦敦南部丹麦山自家花园里开始试着制作一个大模型。选择这个地点部分是因为贝西默不敢去海边试验他的发明。后来他的摇摆交谊厅模型大到可以坐下人了，这样他就可以向同行展示这个想法了。贝西默采用蒸汽机模仿船在海上的运动，据大家说真的能够使摇摆交谊厅保持平衡。最后，这位发明家报告说他的各位工程师同行已宣布他的试验取得圆满成功。

为了监制一艘专门容纳这项发明的跨海峡轮船，很快成立了一家贝西默交谊厅轮船公司。这艘轮船长约107米（350英尺），比当时的许多轮船要大得多。这一长度可以减小轮船前后的颠簸程度，而摇摆交谊厅则可以使船上的其他运动减小到微乎其微。

1872年，这艘船令人尊敬的设计师里德（E.J.Reed，曾任皇家海军的总工程师）给《泰晤士报》写了一封信，

注释

① remedy ['rɛmɪdɪ] n. 补救办法
② neutralise ['njuːtrəˌlaɪz] v. 中和
③ plough [plaʊ] v.犁（地）
④ fixture ['fɪkstʃə] n.（房屋内如浴缸、马桶等的）固定装置
⑤ gilt [gɪlt] adj. 镀金的
⑥ mural ['mjʊərəl] n. 壁画

front and back of the ship. The saloon would help reduce the rest of the motion aboard to virtually nothing.

In a letter to *The Times* in 1872 the ship's respected architect, E.J. Reed, who had once been chief constructor of the Royal Navy, explained how he saw the SS *Bessemer* working:

I do not put her forward as a perfect remedy for sea-sickness in all cases, although I think she will be found a sufficient remedy① in the Straits of Dover. Her advantages seem to me to be that she will be large enough herself to escape all but very small movements as regards lifting bodily and pitching. The moderate pitching which she would otherwise experience will be diminished by the low ends, and what remains of it will scarcely be felt at all in the centre saloon. The rolling of the ship, which is the only remaining movement of importance, will be perfectly neutralised② by Mr. Bessemer's hydraulic arrangements.

In the same year the scientific journal *Nature* agreed that the project was surely on course for success with Bessemer and Reed on board: 'The association of those names is in itself a sufficient guarantee that the idea will be carried into execution with complete security as respects the safety of the passengers and the seaworthiness of the ship, and a full knowledge of the scientific principles involved.'

Convinced that his invention was a masterstroke, Bessemer ploughed③ much of his own money into the company and the construction of the ship. No expense was spared on the fixtures④ and fittings of his Saloon either, which measured 70ft by 30ft wide and 20ft high. It featured extravagant carved oak columns, gilt⑤ panels and hand-painted murals⑥. By 1875 the

解释他是如何看待"贝西默号"工作的。

> 我不认为她是预防各种晕船的完美补救措施，不过我认为她足以在多佛海峡应对晕船。在我看来，她的优点在于空间大，足以避免身体摇摆和颠簸等各种运动。即使有少许颠簸，也能通过矮端板消除，而在中央交谊厅中能感觉到的颠簸则微乎其微了。剩下唯一要紧的运动则是船的起伏，这又可以通过贝西默爵士的液压装置得到完美抵消。

同年，科学杂志《自然》也认为这一项目在贝西默和里德的主持下一定可以成功，写道："把这两个名字放在一起，本身就足以保证这个项目可以顺利实施，既能保证乘客的安全，又能保证船的适航性，而且完全掌握了相关科学原理。"

贝西默认为自己的发明实在了不起，就把大量钱投入了这家公司和造船事业。此外，花在摇摆交谊厅的装饰和配件上的钱也不少。这个家伙长约21米（70英尺），宽约9米（30英尺），高约6米（20英尺），具有奢华的雕刻橡木柱、镀金面板和手绘壁画。1875年，造船厂交付了"贝西默号"，其首次公开航行定于5月8日。

为了安全起见，"贝西默号"于4月在英吉利海峡进行了一次试航。事态并未按计划发展，船在加来试着靠码头时撞上了码头墩，撞坏了一扇明轮。随着公开下水的日益临近，人们开始着急忙慌地修船。这样修船意味着贝西

SS *Bessemer* was delivered from the shipyard and its first public voyage was set for 8 May.

Just to be on the safe side, in April the ship made a trial run across the Channel. It didn't go to plan. The ship had bumped[①] into the pier at Calais trying to dock, wrecking one of its own paddle wheels. With the much-publicised public launch approaching the ship was hastily[②] fixed. These repairs meant Bessemer had no chance to finish adjusting the complex mechanism that controlled his saloon. However, instead of delaying the launch, Bessemer decided that, on this occasion, the saloon would have to be fixed rigidly in place for the trip. He'd have no chance to demonstrate just what it could do but at least he could show off the grandeur[③] of the saloon's interior[④] and get people talking. The mechanism could simply be put fully into action on a later run.

After an uneventful passage across the Channel, with its passengers dutifully admiring the saloon's handiwork, the ship was approaching Calais harbour in clement weather. Suddenly disaster struck. The captain had calmly ordered the ship to be moved in one direction – only for it to lurch in another. Bessemer describes how, to his horror, the ship began heading towards Calais pier once again, only this time the collision[⑤] was to be much worse than the one in April. With a loud, splitting[⑥] sound the ship ploughed into the structure knocking down its timbers[⑦] 'like ninepins'. One passenger later admitted, with some shame, that the disaster was met with a great roar of laughter on board. The great man himself wrote:

I knew what it all meant to me. Those five minutes had made me a poorer man by £34,000; it had deprived[⑧] me of one of the greatest triumphs of a long professional life, and had wrought the loss of the dearly-cherished

注释

① bump [bʌmp] v. 撞上
② hasty ['heɪstɪ] adj. 匆忙的
③ grandeur ['grændʒə] n. 宏伟 壮观
④ interior [ɪn'tɪərɪə] n. 内部
⑤ collision [kə'lɪʒən] n.（文化或 观点的）冲突
⑥ split [splɪt] v. 分开
⑦ timber ['tɪmbə] n.木材
⑧ deprive [dɪ'praɪv] v. 剥夺，使 不能有

默没有机会去完成对控制交谊厅复杂机械的调整。然而为了不推迟下水，贝西默临时决定先把交谊厅紧紧固定起来以应对这次航行。他虽然没有机会展示它可以做什么，但至少可以展示交谊厅内饰的宏伟，供人们来谈论。这套机械可以在之后的一次运行中进行完全展示。

在平安无事地穿过英吉利海峡之后，"贝西默号"在风和日丽的天气下驶向加来港，乘客由衷地钦佩交谊厅的做工精美。突然灾难降临了。船长冷静地指挥船沿一个方向运动，结果却向另一个方向倾斜。后来贝西默说，令他惊讶的是，船再次撞上加来港的码头墩，只是这次比4月的那次严重得多。船发出巨大的撕裂声冲进码头墩，把码头墩的柱子撞成了"九柱戏"一般。一位乘客后来略带羞愧地承认说，这次灾难遭到了船上人员的哄堂大笑。伟大的贝西默自己写道：

> 我知道这对我来说意味着什么。那5分钟使我损失了34000英镑，我成了穷人。它还剥夺了我长期职业生涯中的一项伟大成果，使多年来鼓励我、支持我劳动的宝贵希望成了泡影。我曾经天真地希望能够根除那些往来于英吉利海峡两岸成千上万人的痛苦。

这次事故之后，贝西默坚持认为当时他没有机会完善交谊厅系统和使其通过适当的试验。不过好像是这艘船的船长皮托克（Pittock）发现了，很难在入港所需低速下驾驶"贝西默号"，他可是位在英吉利海峡航行的老水手

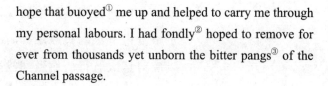

hope that buoyed① me up and helped to carry me through my personal labours. I had fondly② hoped to remove for ever from thousands yet unborn the bitter pangs③ of the Channel passage.

After the event, Bessemer maintained that he'd not had the chance to finesse his saloon system and put it through a proper trial. But it seems that the ship's Captain Pittock, a veteran of the Channel run, had found the craft built to house it difficult to steer at the slow speed needed for entering a harbour. The weight of the Bessemer Saloon in the middle and its heavy engines at either end couldn't have helped. The failure of the voyage made investors panic and in 1876 The Bessemer Saloon Steamboat Company was wound up.

The SS *Bessemer* itself was scrapped, though its architect, E.J. Reed, eventually had the saloon section moved to his own home in Stanley, Kent where it became a private billiard④ room. It survived until the Second World War when a direct hit from a German bomb blew up all that remained of Bessemer's floating folly.

Bessemer's reputation managed to recover from the embarrassment of the episode and history has been kind to him. Today thirteen towns in the USA still bear his name. He might be disappointed that no one since him has successfully come up with a design of ship that entirely can cure seasickness. But he'd probably be thankful that, these days, he would be able to take a flight across the Atlantic to visit all those American towns which share his name, rather than endure agonising⑤ weeks at sea.

了。贝西默交谊厅的重量位于中间，而在其任一端的重型发动机无法发挥作用。航行失败使投资者感到恐慌，而贝西默交谊厅轮船公司在1876年歇业了。

"贝西默号"被拆成一堆，后来设计师里德则把交谊厅部分搬回肯特郡斯坦利的家中，变成一间私人台球室。直到第二次世界大战时，一枚德国炸弹直接击中它，才算彻底结束了贝西默的这个浮动荒唐之物。

贝西默的名声最终从这次事件的困窘中得以恢复，历史对他也算仁慈。今天，美国的13个城镇仍然承载着他的名字。自他以后，再也没人成功地设计出一种能够完全治疗晕船的船，这可能会让他有点失望。不过他应该感到庆幸，如果现在他还活着，就可以坐飞机跨过大西洋去游览那些以他名字命名的美国城镇了，不用忍受在海上数周的煎熬。

ILLUSTRATIONS
插　图

We would like to express our thanks to everyone who has helped with the illustrations in this book. We have made every attempt to trace owners of copyright material and to attribute each correctly. We apologise for any omissions and will be pleased to incorporate missing acknowledgements in any future editions.

我们谨向为本书提供插图的所有人表示感谢。我们尽了最大努力，希望找到版权资料的所有者并予以适当感谢。对于存在的疏漏，我们深表歉意，但愿在今后的版本中补上遗漏的致谢。

编者注

考虑到读者依据原文检索更为方便，故对本书插图提供机构名称及人员姓名未作翻译。

Pigeon-guided missiles　鸽子制导导弹

Image courtesy of the B.F. Skinner Foundation

A sound plan for defence　声波防御计划

Images copyright Peter Spurgeon, www.peterspurgeon.net

The Darien debacle　达连计划的失败

Image courtesy of the University of Glasgow

The lost US state of Transylvania　丢失的美国特兰西瓦尼亚州

Image of Transylvania Convention, public domain

A tax on light and air　光和空气税

Image copyright James Moore

The 'Spruce Goose'　"云杉鹅号"飞机

Image courtesy of the Evergreen Aviation and Space Museum, www.sprucegoose.org

Exploding traffic lights　爆炸的红绿灯

Poster image courtesy of Metropolitan Police Archives Plaque image copyright James Moore

The steam-powered passenger carriage　蒸汽客车

Cugnot's Fardier, Image courtesy of Musée des Arts et Métiers, Paris Steam racing car, The State Archives of Florida

Flying cars　飞行汽车

Images courtesy the Museum of Flight, Seattle, www.museumofflight.org

The atomic automobile　原子能汽车

Images courtesy of The Ford Motor Company

Cincinnati's subway to nowhere　辛辛那提不通车的地铁
Image courtesy of the Cincinnati Museum Center

Is it a train or a plane　它是火车还是飞机
Image courtesy of Information & Archives, East Dunbartonshire Libraries

How the Cape to Cairo railway hit the buffers　好望角至开罗的铁路如何遭遇搁浅
Image from Punch magazine, public domain

British Rail's flying saucer & the 'great space elevator'　英国铁路公司的"飞碟"和大太空升降机

Original patent image courtesy of the European Patent Office, retrieved from the esp@cenet services, ep.espacenet.com

Why the bathyscaphe was the depth of ambition　为什么深海潜水器象征雄心的深度
Image courtesy of Naval History and Heritage Command

SOURCES AND RESOURCES

资料来源

Pigeon-guided missiles　鸽子制导导弹

National Museum of American History Archives.

American Psychologist (Vol. 15, Issue 1, January 1960), pp. 28-37.

BF Skinner Foundation archive.

Psychology Today (14 July 2010).

BBC Online article (8 March 2000).

Daily Telegraph (7 December 2009).

A sound plan for defence　声波防御计划

Richard Scarth, *Echoes From The Sky: A Story of Acoustic Defence* (Hythe Civic Society, 1999).

'Listening For The Enemy', *Cabinet Magazine* (fall/winter 2003).

Mark Denny, *Blip*, *Ping* & *Buzz* (The John Hopkins University Press, 2007).

编者注

考虑到读者依据原文检索更为方便，故对本书资料来源未作翻译，但依据《信息与文献参考文献著录规则》对原书文献著录格式进行了修改，并在能检索到的范围内对原书文献中的缺项信息进行了增补。

The diabolical death ray　残忍的死亡射线

New Scientist (23 December 1976).
Jonathan Foster, *The Death Ray-The Secret Life of Harry Grindell Matthews* (Inventive Publishing, 2009).
Time Magazine (21 April 1924).
Fortean Times (October 2003).
South Wales Echo (30 July 2002).

The misplaced Maginot Line　错位的马其诺防线

Time Magazine (18 January 1932).
Ian Ousby, Occupation: The Ordeal of France (Pimlico, 1997).
The Historian (22 June 2008).

The great 'Panjandrum'　武器中的"大亨"

Brian Johnson, The Secret War (Arrow Press, 1978).
Di James Kiras, *Special operations and strategy: from World War II to the War on Terrorism* (Routledge, 2006).
Atlantic Wall Linear Museum, www.atlanticwall.polimi.it

The Darien debacle　达连计划的失败

Spencer Collection, University of Glasgow.
The Guardian (11 September 2007).
The Scotsman (28 April 2007).
Dr Mike Ibeji, 'The Darien Venture', on www.bbc.co.uk
History Today (1 November 1998).

The lost US state of Transylvania　丢失的美国特兰西瓦尼亚州

Christian G. Fritz, *American Sovereigns: The People and America's Constitutional Tradition Before the Civil War* (Cambridge University Press, 2008).
Michael J. Trinklein, *Lost States: True Stories of Texlahoma, Transylvania, and Other States That Never Made it* (Quirk Books, 2010).

The Columbia Encyclopaedia (Sixth Edition).

John E. Kleber, *The Kentucky Encyclopaedia* (1992).

The French republican calendar 法国共和历

David Ewing Duncan, *The Calendar* (Fourth Estate, 1998).

Matthew John Shaw, *Time and the French Revolution, 1789-Year XVI* (M.Phil thesis, University of York, 2000).

'The Fondation Napoléon', on www.napoleon.org

'Calendars through the ages', on www.webexhibits.org/calendars

Latin Monetary Union 拉丁货币同盟

Kee-Hong Bae and Warren Bailey, *The Latin Monetary Union: some evidence on Europe's failed common currency* (Korea University and Cornell University, 2003).

Luca Einaudi, *European Monetary Unification and the International Gold Standard (1865-1873)* (Oxford University Press, 2001).

Henry Parker Willis, *A history of the Latin Monetary Union: A study of international monetary action* (University of Chicago Press, 1901).

A tax on light and air 光和空气税

Glyn Davies, *A history of money from ancient times to the present day* (University of Wales Press, 2002).

Andrew E. Glantz, 'A Tax on Light and Air: Impact of the Window Duty on Tax Administration and Architecture, 1696-1865', *Penn History Review* (Vol. 15, Issue 2, University of Pennsylvania, 2008).

G. Timmins, 'The History of Longparish', on www.longparish.org.uk/history/windowtax

W.R. Ward, 'Administration of the Window and Assessed Taxes 1696-1798', *The English Historical Review* (Oxford, 1952).

Stephen Dowell, *A History of Taxation and Taxes in England* (Longman Green, 1884).

The international 'hot air' airline 国际"蒸汽"航线

C. Fayette Taylor, 'A review of the evolution of aircraft piston engines', *Smithsonian Annals of Flight* (Vol. 1, No 4, 1971).

Clive Hart, *A Prehistory of flight* (University of California Press, 1985). Russell Naughton, *Hargrave Aviation and Aeromodelling* (Pandora Archive, University of Canberra, 2007).

Chard Museum, Somerset holds many of John Stringfellow's documents, including the Henson and Stringfellow model planes.

London's Science Museum holds copies of the advertisements and other promotional material for the proposed international airline.

The 'Spruce Goose'　"云杉鹅号"飞机

Evergreen Aviation Museum, www.sprucegoose.org

The Columbian (6 May 2003).

Popular Science (September 1945).

Popular Mechanics.

Air Power History (22 September 2007).

The Economist (12 December 1992).

Wood Based Panels International (1 June 2004).

The Press (1 January 2007).

Exploding traffic lights　爆炸的红绿灯

Westminster City Council leaflet and other materials from the Metropolitan Police Historical Archive, West London.

Daily Mail (5 March 1998).

'Minutes of the proceedings' (Institute of Civil Engineers, 1887).

Police Review (29 April 1983).

The steam-powered passenger carriage　蒸汽客车

Di Thomas Kingston Derry and Trevor Illtyd Williams, *A short history of technology: from the earliest times to ad 1900* (Oxford University Press, 1960).

Alain A. Cerf, 'Nicholas Cugnot, Fardier 1770', on www.nicolascugnot.com

Di Lita Epstein, Charles Jaco and Julianne C. Iwersen-Niemann, *The complete idiot's guide to the politics of oil* (Alpha Books, 2003).

Samuels Smiles, *The lives of engineers: George and Robert Stephenson* (Salzwasser-Verlag, 2010).

'The Stanley Steam Engine', on www.stanleymotorcarriage.com

Flying cars　飞行汽车

Life (16 August 1937).
Popular Science (2000).
Times (15 May 2004).
The Seattle Times (5 September 2006).
The Seattle Times (15 July 1990).
New York Times (11 April 2009).
New Scientist (29 May 1999).

The atomic automobile　原子能汽车

Original Publicity Documents, Ford (1958).
'When Dream Cars Collide With Real-World Demands', *New York Times* (7 January 2007).
'Nuclear Powered Passenger Aircraft to Transport Millions', *Times* (27 October 2008).

Cincinnati's subway to nowhere　辛辛那提不通车的地铁

Allen J. Singer, *The Cincinnati Subway* (2003).
Cincinnati Enquirer (24 May 2003).
Cincinnati Enquirer (29 July 2002).
City of Cincinnati government website, http://www.cincinnati-oh. gov/
Cincinnati Examiner (1 June 2010).
Cincinnati Post (23 March 2007).

Is it a train or a plane　它是火车还是飞机

'The George Bennie Railplane System', Brochure, National Archives of Scotland.
Christian Wolmar, *Blood, Iron & Gold* (2009).
William B. Black, *The Bennie Railplane* (East Dunbartonshire Council, 2004).
The Times (13 March 2006).
The Guardian (13 March 2006).
Milngavie Herald (23 August 2006).
The Sunday Times (18 April 2010).

How the Cape to Cairo railway hit the buffers　好望角至开罗的铁路如何遭遇搁浅

Albert A. Hopkins, *Scientific American Reference Book* (A. Russell Bon, 1905).

George Tabor, *The Cape To Cairo Railway and River Routes* (Genta Publications, 2003).

New York Times (8 March 1908).

J.G. Macdonald, *Rhodes – A Life*.

From Russia to America by train　坐火车从俄国到美国

James A. Oliver, *The Bering Strait Crossing* (2006).

Time (13 March 1944).

'Extreme Engineering', *Discovery Channel* (2003).

Popular Science Magazine (December, 2004).

British Rail's flying saucer & the 'great space elevator'　英国铁路公司的"飞碟"和大太空升降机

The Complete Dictionary of Scientific Biography.

Michel Van Pelt, *Space Tethers and Space Elevators* (Praxis Publishing, 2009).

Daily Mail (28 February 1996).

The Guardian (29 April 2006).

Why the bathyscaphe was the depth of ambition　为什么深海潜水器象征雄心的深度

Bill Bryson, *A short history of nearly everything* (Doubleday, 2003).

T.A. Heppenheimer, 'To the bottom of the sea', *Invention and Technology Magazine* (Vol. 8, Issue 1: summer 1992).

Robert Gannon, 'What really happened to the Thresher', *Popular Science* (February 1964).

'The deepest explorers', a website dedicated to Auguste and Jacques Piccard: www.deepestdive.com

Brunel's not so Great Eastern　布鲁内尔名不副实的"大东方号"

George S. Emmerson, 'L.T.C. Rolt and the Great Eastern Affair of Brunel versus Scott Russell', *Technology and Culture* (Vol. 21, No 4, The Johns Hopkins University Press, October 1980).

'Great Eastern: The Launch Of The "Leviathan"', *Mechanics Magazine* (19 December 1857).

Keith Hickman, 'Brunel's "Great Eastern" Steamship, The Launch Fiasco–An Investigation', *Gloucestershire Society for Industrial Archaeology Journal* (2005).

For pictures and more information about the SS *Great Eastern* see www.brunel200.com

Bessemer's anti-seasickness ship   贝西默的防晕船轮船

Sir Henry Bessemer, F.R.S. An Autobiography (1905).

Percy Hethrington Fitzgerald, *Short Works of Percy Hethrington Fitzgerald* (Biblio Bazaar, 2008).

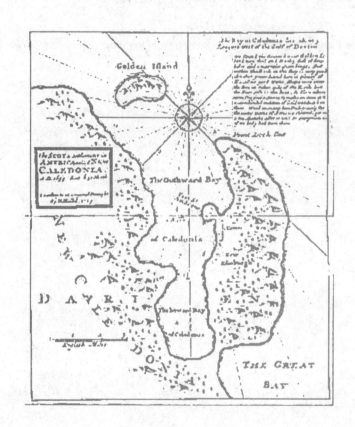

（上册完）